U0306175

Compiled by Chinese Academy of Tropical Agricultural Sciences (CATAS) and Chinese Society for Tropical Crops（CSTC）

A Series of Books for Field Guide to Common Plants in FSM

General Editor：Liu Guodao

Field Guide to Fruits and Vegetables in the Federated States of Micronesia

Editors in Chief：Li Weiming Wang Jinhui

China Agricultural Science and Technology Press

图书在版编目（CIP）数据

密克罗尼西亚联邦果蔬植物图鉴 = Field Guide to Fruits and Vegetables in the Federated States of Micronesia / 李伟明，王金辉主编 . —北京：中国农业科学技术出版社，2021.5

（密克罗尼西亚常见植物图鉴系列丛书 / 刘国道主编）

ISBN 978-7-5116-5291-1

Ⅰ . ①密… Ⅱ . ①李… ②王… Ⅲ . ①水果—种质资源—密克罗尼西亚联邦—图集②蔬菜—种质资源—密克罗尼西亚联邦—图集 Ⅳ . ① S660.24-64 ② S630.24-64

中国版本图书馆 CIP 数据核字（2021）第 069036 号

责任编辑 徐定娜
责任校对 贾海霞
责任印制 姜义伟　王思文

出 版 者 中国农业科学技术出版社
北京市中关村南大街 12 号　邮编：100081
电　　话 （010）82105169（编辑室）（010）82109702（发行部）
（010）82109709（读者服务部）
传　　真 （010）82109707
网　　址 http://www.castp.cn
发　　行 各地新华书店
印 刷 者 北京科信印刷有限公司
开　　本 787 mm × 1 092 mm　1 /16
印　　张 5.75
字　　数 268 千字
版　　次 2021 年 5 月第 1 版　2021 年 5 月第 1 次印刷
定　　价 108.00 元

About the Author

Dr. Liu Guodao, born in June 1963 in Tengchong City, Yunnan province, is the incumbent Vice President of Chinese Academy of Tropical Agricultural Sciences (CATAS). Being a professor and PhD tutor, he also serves as the Director-General of the China-Republic of the Congo Agricultural Technology Demonstration Center.

In 2007, he was granted with his PhD degree from the South China University of Tropical Agriculture, majoring in Crop Cultivation and Farming.

Apart from focusing on the work of CATAS, he also acts as a tutor of PhD candidates at Hainan University, Member of the Steering Committee of the FAO Tropical Agriculture Platform (TAP), Council Member of the International Rubber Research and Development Board (IRRDB), Chairman of the Chinese Society for Tropical Crops, Chairman of the Botanical Society of Hainan, Executive Director of the Chinese Grassland Society and Deputy Director of the National Committee for the Examination and Approval of Forage Varieties and the National Committee for the Examination and Approval of Tropical Crop Varieties.

He has long been engaged in the research of tropical forage. He has presided over 30 national, provincial and ministerial-level projects: namely the "National Project on Key Basic Research (973 Program)" and international cooperation projects of the Ministry of Science and Technology, projects of the National Natural Science Foundation of China, projects of the International Center for Tropical Agriculture in Colombia and a bunch of projects sponsored by the Ministry of Agriculture and Rural Affairs (MARA) including the Talent Support Project, the "948" Program and the Infrastructure Project and Special

Scientific Research Projects of Public Welfare Industry.

He has published more than 300 monographs in domestic and international journals such as "New Phytologist" "Journal of Experimental Botany" "The Rangeland Journal" "Acta Prataculturae Sinica" "Acta Agrestia Sinica" "Chinese Journal of Tropical Crops", among which there nearly 20 were being included in the SCI database. Besides, he has compiled over 10 monographs, encompassing "Poaceae Plants in Hainan" "Cyperaceae Plants in Hainan" "Forage Plants in Hainan" "Germplasm Resources of Tropical Crops" "Germplasm Resources of Tropical Forage Plants" "Seeds of Tropical Forage Plants" "Chinese Tropical Forage Plant Resources". As the chief editor, he came out a textbook-*Tropical Forage Cultivation*, and two series of books-*Practical Techniques for Animal Husbandry in South China Agricultural Regions* (19 volumes) and *Practical Techniques for Chinese Tropical Agriculture "Going Global"* (16 volumes).

He has won more than 20 provincial-level and ministerial-level science and technology awards. They are the Team Award, the Popular Science Award and the First Prize of the MARA China Agricultural Science and Technology Award, the Special Prize of Hainan Natural Science Award, the First Prize of the Hainan Science and Technology Progress Award and the First Prize of Hainan Science and Technology Achievement Transformation Award.

He developed 23 new forage varieties including Reyan No. 4 King grass. He was granted with 6 patents of invention and 10 utility models by national patent authorities. He is an Outstanding Contributor in Hainan province and a Special Government Allowance Expert of the State Council.

Below are the awards he has won over the years: in 2020, "the Ho Leung Ho Lee Foundation Award for Science and Technology Innovation"; in 2018, "the High-Level Talent of Hainan province" "the Outstanding Talent of Hainan province" "the Hainan Science and Technology Figure"; in 2015, Team Award of "the China Agricultural Science and Technology Award" by the Ministry of Agriculture; in 2012, "the National Outstanding Agricultural Talents Prize" awarded by the Ministry of Agriculture and as team leader of the team award: "the Ministry of Agriculture Innovation Team" (focusing on the research of Tropical forage germplasm innovation and utilization); in 2010, the first-level candidate of the "515 Talent Project" in Hainan province; in 2005, "the Outstanding Talent of Hainan province"; in 2004, the first group of national-level candidates for the "New Century Talents Project" "the 4th Hainan Youth Science and Technology Award" "the 4th Hainan Youth May 4th Medal" "the 8th China Youth Science and Technology Award" "the Hainan Provincial International Science and Technology Cooperation Contribution Award"; in 2003, "a Cross-Century Outstanding Talent" awarded by the Ministry of Education; In 2001, "the 7th China Youth Science and Technology Award" of Chinese Association of Agricultural Science Societies, "the National Advanced Worker of Agricultural Science and Technology"; in 1993, "the Award for Talents with Outstanding Contributions after Returning to China" by the State Administration of Foreign Experts Affairs.

Dr. Li Weiming, born in December 1983 in Guiping City, Guangxi province, is an Associate Professor in the South Subtropical Crops Research Institute of Chinese Academy of Tropical Agricultural Sciences, mainly engaged in the germplasm resources collection and breeding of horticultural plant. In the past few years, he focused on the research of the taxonomy, evaluation and hybridization of wild banana cultivars, and has developed a high-efficient technique system for banana cross-breeding by which certain elite new varieties were bred. He has presided over a number of projects such as the project of the National Natural Science Foundation of China and the major project of the Tibet Natural Science Foundation of China. He has published 16 papers in "Functional & Integrative Genomics" "Plant Pathology" "Acta Horticulturae Sinica" and other journals. The monographs he edited, such as the "Germplasm Resources of Fruit Trees" and the "Color Illustration for Banana Cultivation with Good Quality and High Yield". In 2020, he was awarded the title of "High-Level Talent" by the Zhanjiang Municipal government, Guangdong province.

A Series of Books for Field Guide to Common Plants in FSM

General Editor: Liu Guodao

Field Guide to Fruits and Vegetables in the Federated States of Micronesia

Editorial Board

Editors in chief:

Li Weiming   Wang Jinhui

Associate editors in chief:

Zhang Xue   Li Xiaoxia   Yin Xinxing

Members (in alphabet order of surname):

Fan Haikuo   Gong Shufang   Hao Chaoyun   Huang Guixiu

Liu Daozhen   Li Weiming   Li Xiaoxia   Tang Qinghua

Wang Yuanyuan   Wang Qinglong   Wang Jinhui   Yang Guangsui

Yang Hubiao   You Wen   Yin Xinxing   Zheng Xiaowei

Zhang Xue

Photographers:

Yang Hubiao   Wang Qinglong   Hao Chaoyun   Huang Guixiu

Tang Qinghua   Li Weiming

Translator:

Li Yawei

The President
Palikir, Pohnpei
Federated States of Micronesia

FOREWORD

It is with great pleasure that I present this publication, "Agriculture Guideline Booklet" to the people of the Federated States of Micronesia (FSM).

The Agriculture Guideline Booklet is intended to strengthen the FSM Agriculture Sector by providing farmers and families the latest information that can be used by all in our communities to practice sound agricultural practices and to support and strengthen our local, state and national policies in food security. I am confident that the comprehensive notes, tools and data provided in the guideline booklets will be of great value to our economic development sector.

Special Appreciation is extended to the Government of the People's Republic of China, mostly the Chinese Academy of Tropical Agricultural Sciences (CATAS) for assisting the Government of the FSM especially our sisters' island states in publishing books for agricultural production. Your generous assistance in providing the practical farming techniques in agriculture will make the people of the FSM more agriculturally productive.

I would also like to thank our key staff of the National Government, Department of Resources and Development, the states' agriculture and forestry divisions and all relevant partners and stakeholders for their kind assistance and support extended to the team of Scientists and experts from CATAS during their extensive visit and work done in the FSM in 2018.

We look forward to a mutually beneficial partnership.

Sincerely,

David W. Panuelo
President

Preface

Claiming waters of over 3,000 square kilometers, the vast area where Pacific island countries nestle is home to more than 10,000 islands. Its location at the intersection of the east-west and north-south main traffic artery of Pacific wins itself geo-strategic significance. There are rich natural resources such as agricultural and mineral resources, oil and gas here. The relationship between the Federated States of Micronesia (hereinafter referred to as FSM) and China ushered in a new era in 2014 when Xi Jinping, President of China, and the leader of FSM decided to establish a strategic partnership on the basis of mutual respect and common development. Mr Christian, President of FSM, took a successful visit to China in March 2017 during which a consensus had been reached between the leaders that the traditional relationship should be deepened and pragmatic cooperation (especially in agriculture) be strengthened. This visit pointed out the direction for the development of relationship between the two countries. In November 2018, President Xi visited Papua New Guinea and in a collective meeting met 8 leaders of Pacific Island countries (with whom China has established diplomatic relation). China elevated the relationship between the countries into a comprehensive and strategic one on the basis of mutual respect and common development, a sign foreseeing a brand new prospect of cooperation.

The government of China launched a project aimed at assisting FSM in setting up demonstration farms in 1998. Until now, China has completed 10 agricultural technology cooperation projects. To answer the request of the government of FSM, Chinese Academy of Tropical Agricultural Sciences (hereinafter referred to as CATAS), directly affiliated with the

Ministry of Agriculture and Rural Affairs of China, was elected by the government of China to carry out training courses on agricultural technology in FSM during 2017—2018. The fruitful outcome is an output of training 125 agricultural backbone technicians and a series of popular science books which are entitled "Field Guide to Forages in the Federated States of Micronesia" "Field Guide to Flowers and Ornamental Plants in the Federated States of Micronesia" "Field Guide to Medicinal Plants in the Federated States of Micronesia" "Field Guide to Fruits and Vegetables in the Federated States of Micronesia" "Coconut Germplasm Resources in the Federated States of Micronesia" and "Field Guide to Plant Diseases, Insect Pests and Weeds in the Federated States of Micronesia".

In these books, 492 accessions of germplasm resources such as coconut, fruits, vegetables, flowers, forages, medical plants, and pests and weeds are systematically elaborated with profuse inclusion of pictures. They are rare and precious references to the agricultural resources in FSM, as well as a heart-winning project among China's aids to FSM.

Upon the notable moment of China-Pacific Island Countries Agriculture Ministers Meeting, I would like to send my sincere respect and congratulation to the experts of CATAS and friends from FSM who have contributed remarkably to the production of these books. I am firmly convinced that the exchange between the two countries on agriculture, culture and education will be much closer under the background of the publication of these books and Nadi Declaration of China-Pacific Island Countries Agriculture Ministers Meeting, and that more fruitful results will come about. I also believe that the team of experts in tropical agriculture mainly from the CATAS will make a greater contribution to closer connection in agricultural development strategies and plans between China and FSM, and closer cooperation in exchanges and capacity-building of agriculture staffs, in agricultural science and technology for the development of agriculture of both countries, in agricultural investment and trade, in facilitating FSM to expand industry chain and value chain of agriculture, etc.

Qu Dongyu

Director General

Food and Agriculture Organization of the United Nations

July 23, 2019

Located in the northern and central Pacific region, the Federated States of Micronesia (FSM) is an important hub connecting Asia and America. Micronesia has a large sea area, rich marine resources, good ecological environment, and unique traditional culture.

In the past 30 years since the establishment of diplomatic relations between China and FSM, cooperation in diverse fields at various levels has been further developed. Since the 18th National Congress of the Communist Party of China, under the guidance of Xi Jinping's thoughts on diplomacy, China has adhered to the fine diplomatic tradition of treating all countries as equals, adhered to the principle of upholding justice while pursuing shared interests and the principle of sincerity, real results, affinity, and good faith, and made historic achievements in the development of P.R. China-FSM relations.

The Chinese government attaches great importance to P.R. China-FSM relations and always sees FSM as a good friend and a good partner in the Pacific island region. In 2014, President Xi Jinping and the leader of the FSM made the decision to build a strategic partnership featuring mutual respect and common development, opening a new chapter of P.R. China-FSM relations. In 2017, FSM President Peter Christian made a successful visit to China. President Xi Jinping and President Christian reached broad consensuses on deepening the traditional friendship between the two countries and expanding practical cooperation between the two sides, and thus further promoted P.R. China-FSM relations. In 2018, Chinese President Xi Jinping and Micronesian President Peter Christian had a successful meeting again in PNG and made significant achievements, deciding to upgrade P.R. China-FSM relations to a new stage of Comprehensive Strategic Partnership, thus charting the course for future long-term development of P.R. China-FSM relations.

In 1998, the Chinese government implemented the P.R. China-FSM demonstration farm project in FSM. Ten agricultural technology cooperation projects have been completed, which has become the "golden signboard" for China's aid to FSM. From 2017 to 2018, the Chinese Academy of Tropical Agricultural Sciences (CATAS), directly affiliated with the Ministry of Agriculture and Rural Affairs, conducted a month-long technical training on pest control of coconut trees in FSM at the request of the Government of FSM. 125 agricultural managers, technical personnel and growers were trained in Yap, Chuuk, Kosrae and Pohnpei, and the biological control technology demonstration of the major dangerous pest, Coconut Leaf Beetle, was carried out. At the same time, the experts took advantage of the spare time of the training course and spared no effort to carry out the preliminary evaluation of the investigation and utilization of agricultural resources, such as coconut, areca nut, fruit tree, flower, forage, medicinal plant, melon and vegetable, crop disease, insect pest and weed diseases, in the field in conjunction with Department of Resources and Development of FSM and the vast number of trainees, organized and compiled a series of popular science books, such as "Field Guide to Forages in the Federated States of Micronesia" "Field Guide to Flowers and Ornamental Plants in the Federated States of Micronesia" "Field Guide to Medicinal Plants in the Federated States of Micronesia" "Field Guide to Fruits and Vegetables in the Federated States of Micronesia" "Coconut Germplasm Resources in the Federated States of Micronesia" and "Field Guide to Plant Diseases, Insect Pests and Weeds in the Federated States of Micronesia".

The book introduces 37 kinds of coconut germplasm resources, 60 kinds of fruits and vegetables, 91 kinds of angiosperm flowers as well as 13 kinds of ornamental pteridophytes, 100 kinds of forage plants, 117 kinds of medicinal plants, 74 kinds of crop diseases, pests and weed diseases, in an easy-to-understand manner. It is a rare agricultural resource illustration in FSM. This series of books is not only suitable for the scientific and educational workers of FSM, but also it is a valuable reference book for industry managers, students, growers and all other people who are interested in the agricultural resources of FSM.

This series is of great significance for it is published on the occasion of the 30th anniversary of the establishment of diplomatic relations between the People's Republic of

China and FSM. Here, I would like to pay tribute to the experts from CATAS and the friends in FSM who have made outstanding contributions to this series of books. I congratulate and thank all the participants in this series for their hard and excellent work. I firmly believe that based on this series of books, the agricultural and cultural exchanges between China and FSM will get closer with each passing day, and better results will be achieved more quickly. At the same time, I firmly believe that the Chinese Tropical Agricultural Research Team, with CATAS as its main force, will bring new vigour and make new contributions to promoting the in-depth development of the strategic partnership between the People's Republic of China and the Federated States of Micronesia, strengthening solidarity and cooperation between P.R. China and the developing countries, and the P.R. China-FSM joint pursuit of the Belt and Road initiative and building a community with a shared future for the humanity.

黄峥

Ambassador Extraordinary & Plenipotentiary of
the People's Republic of China to
the Federated States of Micronesia
May 23, 2019

Foreword

The Federated States of Micronesia (hereinafter referred to as FSM), an island state located on the Caroline Islands that stands at the northwest of Pacific Ocean, is one of the three Pacific Islands Regiments. With the characteristics of typical tropical rainforest climate, FSM is sunny, humid and rainy all the year round. Although the temperature is relatively stable in FSM over the year, there is a high diurnal difference with cooler night. The annual precipitation in FSM varies from 4,400 to 5,000 mm, with an average annual temperature of 26 to 28℃ . The interaction of adequate rain and heat and isolation from land has brought about countless local species, namely the vegetation diversity and species diversity. At the same time, the sparse population and rich natural resources let local inhabitants be free of growing food. The whole country is basically in a state of wilderness. Local and imported species are well reserved under such well-maintained environment.

Previous to the Second World War, people in FSM generally took starch stuff such as breadfruit, banana and taro as their staple food. They also enjoy eating sea food and fruit. Due to the healthy diet, local residents rarely suffered from malnutrition, diabetes or obesity. According to a survey conducted by American Navy at the end of the Second World War, local residents are unanimously strong and vigorous. However, the situation is entirely different at present. Over half of the adults fall victim to obesity. Night blindness and respiratory tract infections among children, and cardiovascular disease and cancer among the adults caused by vitamin A deficiency have emerged. This woeful scene is because they eat more introduced rice, flour, meat and milk but less fruit and vegetables.

Fruit and vegetables provide the majority of vitamins necessary to human body. It is therefore of great significance to include fruit and vegetable to upgrade diet structure and improve health. During 2017—2018, Chinese Academy of Tropical Agricultural Sciences sent some scientists to carry out assistance projects in FSM, including Training course on coconut diseases control technologies. A survey of fruit trees and vegetables were made in several FSM islands at the same time with a view of including more fruit and vegetable in their diet.

Sixty species of fruit trees, melon plants and vegetables commonly seen in FSM have been recorded in this book, and each includes one or more varieties. In order to clearly show the distinctive features, one or more pictures are enclosed to the description of each species. These fascinating photos can promote the secretion of saliva. Scientific name, characters and living conditions of each plant are described, and basic information such as its English name, local name and distributed region are provided. The book is a good choice for fruits & vegetables fans, nature lover and common folks (especially local inhabitants in FSM) who likes reading popular science books. It is also a notable book for gardeners and reference book and textbook for researchers.We need badly add that we would be very grateful if any reader can draw our attention to errors and omissions which might be corrected later in our subsequent version.

刘国道

General Editor

Vice President of Chinese Academy of Tropical Agricultural Sciences

March 22, 2019

Contents

Fruits

Vegetables

Fruits

Banana

Latin Name: ***Musa spp*** **Lour.**

English Name: Banana

Plant grows in clump with corms. The height of dwarf banana (Maximum 3.5 m) generally is less than 2 m whereas the giant banana varies from 4–5 m. Pseudostem is universally dark green with black spot, coated with white powder especially at the upper part. The leaf is (1.5) 2–2.2 (2.5) m long, 60–70 (85) cm wide, symmetrical on both sides, obtuse at apex, subrounded at base; upper surface dark green, not coated with white powder; lower surfacelight green, coated with white powder. Petiole is generally less than 30 cm, short, thick; leaf wings conspicuous, open, with brown-red or glossy red margin. Spikes are pendulous; rachis densely pubescent. Bracts are purple red outside, covered with white powder, dark red inside; base is slightly pale, glossy; the bracts of male flowers do not fall off, with two rows of flowers each bract. Flowers tend to be milky white or slightly purple. Free perianth segments are subrounded, entire at margin, conically acute at apex. Connate perianth segments consists of lobules on both sides of the central segment, and lobules are about half the length of the central lobe. The largest fruit bunch has 360 fingers, weighing up to 32 kg. Generally a fruit bunch has 8–10 hands and about 150–200 fingers. Fingers are curved, slightly arched, grow upward and erect when young, gradually straight after mature, 12–30 cm long, 3.4–3.8 cm in diameter, ovate with prominent 4–5 angles, narrow at apex, short pedicellate, skin green. Fruit skin is yellowish when forced to ripen at a high temperature , turns from green to yellow with black spots on the surface when forced at low temperature. Flesh is soft, yellowish white, sweet, seedless and very fragrant. Sword suckers, suckers with sword-like leaves, are about 50 cm high, robust at base, purplish red tinged with grayish green, with prominent large black spots, slender at upper part; leaves narrowly long, growing upward, abaxially covered with a thick layer of white powder.

Distributed in: Pohnpei, Yap, Chuuk, Kosrae

Bread Fruit

Latin Name: ***Artocarpus incisus*** **(Thunb.) L.f.**

English Name: Bread fruit

Yap Name: Thow

Chuuk Name: Mai

An evergreen tree, 10–15 m high. Bark is grayish brown, thick. Leaves are large, alternate, thick, leathery, ovate to ovate-elliptical, 10–50 cm long; mature leaves are pinnately parted, mostly 3–8 pinnately parted ; segments lanceolate acuminate at apex, adaxially glabrous and , dark green, abaxially pale green , entire at margin. Lateral veins are about 10 pairs. Petiole is 8–12 cm long. Stipules are large, lanceolate or broadly lanceolate, 10–25 cm long, yellowish green, gray or brown appressed pubescent. Inflorescence is solitary, axillary Male inflorescence is yellow, narrowly cylindrical to narrowly ellipsoid or rod-shaped, 7–30 (–40) cm long; calyx is tubular, pubescent, 2 lobed at the upper part; lobes are lanceolate; stamen 1; anthers are elliptic. Female calyx is tubular; ovary ovoid; style long; stigma 2 lobed. Syncarp is obovoid or subglobose, 15–30 cm long, 8–15 cm in diameter with a ratio of length to width 1–4, green to yellow, covered with round tuberculate protuberance. When mature the fruit is brown to black with soft pericarp and milky fleshy calyx in the mesocarp. Stone is ellipsoid to conical, about 25 mm in diameter. The cultivated varieties have few or no stones.

Distributed in: Pohnpei, Yap, Chuuk, Kosrae

Wax Jambo

Latin Name: ***Syzygium samarangense*** **(Bl.) Merr. et Perry**

English Name: Mountain apple (in Hawaii), wax apple, Malay apple

Yap Name: Arfath

Chuuk Name: Amot apple, Mountain apple

Trees, 12 m tall, twigs flattened. Leaves are thin, coriaceous, elliptic to oblong, 10–22 cm long, 5–8 cm wide, narrow, round or subcordate at base, slightly acute or obtuse at apex, abaxially glandular, adaxially yellowish brown when dry; lateral veins are 14–19 on each side of the midvein obliquely upward at an angle of 45 degrees, 6–10 mm apart, obviously reticulate; intramarginal veins about 5 mm away from the margin are conspicuous, with an additional intramarginal vein about 1.5 mm away from the margin. Petioles are very short, not longer than 4 mm, sometimes nearly sessile. Racemes are terminal or axillary, 5–6 cm long, several–flowered. Flowers are white. Pedicels are about 5 mm. Calyx tube is obconical, 7–8 mm long, 6–7 mm wide; teeth 4, semicircular, 4 mm long, 8 mm wide. Stamens are numerous, about 1.5 cm long; style 2.5–3 cm long. Fruit are pyriform or conical, fleshy, glossy, dark red, 4–5 cm long, with concave at apex, with persistent fleshy sepals and one seed. Flowering is from March to April, and fruit mature from May to June.

Distributed in: Pohnpei, Yap, Chuuk, Kosrae

Papaya

Latin Name: ***Carica papaya*** **L.**

English Name: Papaya

Yap Name: Baibaai, Pawpaw

Papaya is a herbaceous arborescent plant, 8–10 m high, laticiferous. Stems are unbranched or sometimes branched at the site of injury, with spirally arranged scars. Leaves are large, up to 60 cm in diameter, nearly peltate, clustered at the growing tip of the stem, often 5–9 parted, pinnatly lobed on each segment each lobe. Petiole is hollowe, 60–100 cm long. Flowers are male, female or perfect. Some varieties occasionally produce hermaphrodite or female flowers on male plants and bear fruits. Sometimes a few male flowers appear on female plants. Papaya plants are male, female or bisexual. Male flowers are arranged in panicle inflorescence which is 1 m long, pendulous, sessile; sepals confluent at base; corolla creamy yellow; corolla tube slender, 1.6–2.5 cm long; corolla segments 5, lanceolate, 1.8 cm long, 4.5 mm wide; stamens 10, 5 long, 5 short, almost with no filaments for the short ones, with white filament and white pubescence for the long ones. Ovary is degenerated. Female flowers are solitary or in few-flowered corymbose cymes, borne in the leaf axils; pedicels short or nearly sessile; sepals 5, about 1 cm long, connate below the middle; corolla segments 5 free, cream yellow or yellowish white, oblong or lanceolate, 5–6.2 cm long and 1.2–2 cm wide; ovary superiou, ovoid, and sessile; stigmas 5, several lobed, nearly limbriate. Perfect flowers have 5 stamens, borne on the very short corolla tube near the base of ovary, or 10 stamens on the longer corolla tube in 2 whorls; corolla tube is 1.9–2.5 cm long, corolla lobes oblong, about 2.8 cm long and 9 mm wide; ovary is smaller than that of the female plant. The fruit is a berry, fleshy, orange or yellow when ripe elongated globose, obovate elongated globose, pyriform or subglobose, 10–30 cm long or longer; flesh soft, juicy, sweet; Seeds are plentiful, ovoid, black when ripe, fleshy in outer layer of seed coat, lignified in the inner layer of seed coat, wrinkled. Flowering and fruiting are all year round.

Distributed in: Pohnpei, Yap, Chuuk, Kosrae

Guava

Latin Name: ***Psidium guajava*** **L.**

English Name: Guava

Yap Name: Abas, Kuava

Tree, up to 13 m tall; bark gray, smooth, flaky; twigs are angled, puberscent. Leaves are leathery, oblong to elliptic, 6–12 cm long and 3.5–6 cm wide, acute or obtuse at apex, rounded near base, rough adaxially, pubescent abaxially; lateral veins 12–15 on each side of the midveins, often impressed, obviously reticulate; Petiole is 5 mm long. Inflorescences are solitary or 2–3 flowered in cymes; calyx tube pyriform, or campenulate, 5 mm long, pubescent; calyx cap subrounded, 7–8 mm long, irregularly dehiscent; petals 1–1.4 cm long, white; stamens 6–9 mm long. Ovary is inferior, connate to calyx; style and stamen are equal in length. Fruit is a berry, globose, oval or pyriform, 3–8 cm long, with persistent calyx segments at apex; flesh white or yellow; placenta well developed, fleshy, reddish; seeds plentiful.

Distributed in: Pohnpei, Yap, Chuuk, Kosrae

Citrus

Latin Name: ***Citrus*** **spp.**

Tree, branches are numerous, expanding or slightly pendulous, few thorned. Leaves are unifoliate compound; leaf wings are usually narrow or only of remnant at base with which petiole is winged; leaf blade lanciolate, elliptic or broadly ovate, various in size, usually concave at apex; midvein furcate from the base of leaf blade to the place near the concave apex; margin usually obtusely dentate or crenulate at least in the upper part, rarely entire. Inflorescences are solitary or in a fascicle of 2–3 flowers; calyx irregularly 3–5 lobed; petals usually less than 1.5 cm long; stamens 20–25; style long, slender; stigma clavate. Fruits are variable in shape, usually oblate to subglobose. Rind (pericarp) is very thin and smooth, or thick and coarse, pale yellow, red or carmine, easy or slightly easy to peel. Pith (the stringy central portion and membranous walls) plentiful or few, free, usually tender; central pith large, usually hollow, rarely stuffed. Sarcocarp segments are 7–14, seldom more; walls thin or slightly thick, soft, tender or rather tough; pulp vesicles usually spindle-shaped, short, swollen, rarely slender and long; pulp (sarcocarp) sweet to acidic, and sometimes bitter. Seeds are few to many, rarely absent, usually ovate, acute at apex, rounded at base. The cotyledons are dark green, pale green or sometimes nearly milky white; chalazal purple, polyembryonic, rarely single embryo. Flowering is from April to May and fruiting from October to December.

Distributed in: Pohnpei, Yap, Chuuk, Kosrae

Pineapple

Latin Name: ***Ananas comosus*** **(Linn.) Merr.**

English Name: Pineapple

Stems are short. Leaves are numerous, arranged in a rosette around the stem, sword-like, 40–90 cm long and 4–7 cm wide, acuminate at apex, entire or sharply serrate at margin, abaxially green, adaxially pinkish green, usually tinged with brownish red at margin and apex; top leaves borne on the top of the inflorescence small, usually red. Inflorescence emerge from the center of the rosette of flowers, looks like a pine cone, is 6–8 cm long, enlarged when fruiting; bract is green at base, pale red in the upper part, triangularly ovate; sepals are ovate, fleshy, red at apex, 1 cm long ; petals are oblong, acute at apex, about 2 cm long, purplish red at upper part, white at lower part. Syncarp is fleshy, more than 15 cm long. Flowering is from summer to winter.

Distributed in: Pohnpei, Yap, Chuuk, Kosrae

Soursop

Latin Name: ***Annona muricata*** **L.**

English Name: Soursop

Yap Name: Sausau

Evergreen tree, up to 8 m tall. Bark is coarse. Leaves are papery, obovately oblong to elliptic, 5–18 cm long and 2–7 cm wide, acute or obtuse at apex, broadly cuneate or rounded at base, adaxially emerald green and lustrous, abaxially pale green, glabrous on both surfaces. Lateral veins are 8–13 on each side of the midrib, slightly protrudent on both surfaces; reticulate veinlets close to margin. Flower buds are ovate; flowers pale yellow, 3.8 cm long, long as diameter or slightly wider. Sepals are ovate-elliptic, persistent, about 5 mm long. Outer petals are thick, broadly triangular, 2.5–5 cm long, acute to obtuse at apex, with small red protrusions at the inner side of the base, sessile, valvate; inner petals are slightly thin, ovate-elliptic, 2–3.5 cm long, obtuse at apex, clawed at base, short stalked, imbricate. Stamens are about 4 mm long; filaments fleshy, connective dilated. Carpels are 5 mm long, white pubescent. Syncarp is ovate, 10–35 cm long, 7–15 cm in diameter, dark green; spines curved downward when young, caduceus, of remnant with small protuberances when the fruit is mature. Flesh is slightly sour, juicy, white. Seeds are numerous, kidney-shaped, 1.7 cm long and 1 cm wide, brownish yellow. Flowering is from April to July and fruiting from July to March.

Distributed in: Pohnpei, Yap, Chuuk, Kosrae

Mango

Latin Name: ***Mangifera indica*** **L.**

English Name: Mango

Yap Name: Manga

Chuuk Name: Kangit; Manko

Pohnpei Name: Kangit

Evergreen tree, 10–20 m tall. Bark glabrous, grayish brown, brown in branchlets. Leaves are thin and leathery, often aggregated at the end of twigs, variable in shape and size, usually oblong or oblong lanceolate, 12–30 cm long and 3.5–6.5 cm wide, slightly glossy adaxially; apex acuminate or acute; base cuneate or subcircular; margin glabrous, crisped; lateral veins 20–25 on each side of the midrib, obliquely ascending, protuberant on both surfaces, not obviously reticulate. Petiole is 2–6 cm long, grooved at upper part, dilated at base. Panicles are 20–35 cm long, densely flowered, grayish-yellow puberulent; branches spread, 6–15 cm long at base. Bracts lanceolate, 1.5 mm long, puberulent. Flowers small, polygamous, yellow or pale yellow. Pedicels are nodular, 1.5–3 mm long. Sepals are ovate-lanceolate, 2.5–3 mm long, 1.5 mm wide, acuminate, abaxially puberulent, ciliated at margin. Petals are glabrous, oblong or oblong-lanceolate, 3.5–4 mm long, 1.5 mm wide, adaxially veined with 3–5 brown protuberances, curled at anthesis. Floral disk is enlarged, fleshy, 5 lobed. Fertiletamen is 1, about 2.5 mm long; anthers ovate. Stamnodes are 3–4 with or without very short filaments and tuberculate anther primordia. Ovary is glabrous, obliquely ovate, 1.5 mm in diameter. Style is nearly terminal, 2.5 mm long. Drupes are large, kidney-shaped (cultivars vary greatly in shape and size) , flat, 5–10 cm long, 3–4.5 cm wide, yellow when ripe; mesocarp thick, fleshy, bright yellow, sweet. Seeds are hard.

Distributed in: Pohnpei, Yap, Chuuk, Kosrae

Star Fruit

Latin Name: ***Averrhoa carambola*** **L.**

English Name: Star Fruit, Country Gooseberry

Pohnpei Name: Ansu

Tree, up to 12 m tall, with many branches. Bark is dark gray, pale yellow in the inner side, brown when dry, sweet and astringent. Leaves are alternate, odd-pinnately compound, 10–20 cm long; leaflets 5–13, ovate or elliptic, 3–7 cm long and 2–3.5 cm wide, entire at margin, acuminate at apex, round and tilted in one side at base, dark green adaxially, pale green abaxially, sparsely pubescent or glabrous, short petiolulate. Flowers are small, slightly fragrant, numerous, clustered in cymes or panicles, axillary or on the stems of branches, crimson at floral branches and buds. Sepals 5, imbricate, 5 mm long, slender cupulate. Petals are slightly curled towards the lower side, 8–10 mm long and 3–4 mm wide, purplishred abaxially; margin pale, sometimes pink or white. Stamens are 5–10; styles 5. Ovaries 5, with many ovules each ovary. Berry is fleshy, pendulous, 5-ribbed, rarely 3 or 6-ribbed, stellate in cross section, 5–8 cm long, pale green or waxy yellow, sometimes tinged with dark red. Seeds are dark brown. Flowering is from April to December and fruiting from July to December.

Distributed in: Pohnpei, Chuuk

Garlic Pear

Latin Name: ***Crataeva religiosa*** **G. Forst**

English Name: Garlic pear

Yap Name: Abiuuch; Abiich

Chuuk Name: Abuts

Pohnpei Name: Apoot

Small tree. Branches are distinct in lenticels. Leaves are trifoliate; leaflets ovate, acute at apex, rounded at base, 10 cm long, 4 cm wide. Petioles are 10–15 cm long; petiolules short, 2–3 mm long. Flowers terminal or axillary, aggregated in cymes. Petals are white, ovate, unequal in size, petiolate, about 3.5 cm long at most. Stamens are numerous, unequal in length, longer than petals. Fruit is ovoid to cylindrical, 6 to 15 cm long. Seeds are embedded in fleshy berries, usually absent in large fruits.

Distributed in: Pohnpei, Yap

Passion Fruit

Latin Name: ***Passiflora caerulea*** **L.**

English Name: Passion fruit

Yap Name: Tumatis

A herbaceous perennial vine. Stems are cylindrical, slightly angled, glabrous, slightly white pruinose. Leaves are papery, 5–7 cm long, 6–8 cm wide, cordate at base and palmately 5 lobed; central lobe is oval-oblong; lobes in both sides slightly smaller, glabrous, entire at margin. Petiole is 2–3 cm long, with 2–4 small glands in the middle. Stipules are large, kidney-shaped, amplexicaul, up to 1.2 cm long, undulate at margin. Cymes are degenerated, only 1 flowered, opposite to tendrils. Flowers are large, pale green, up to 6–8 (10) cm in diameter. Pedicels are 3–4 cm long. Bracts are ovate, 3 cm long, entire at margin. Sepals are 5, 3–4.5 cm long, adaxially pale green, abaxially greenish white, with an angular appendage at the abaxial top. Petals are 5, pale green, subequal to sepals. Outer corona lobes are in 3 series, filiform; lobes in outer and middle series 1–1.5 cm long, sky blue at apex, white at middle, purple red at base; lobes in inner series are filiform, 1–2 mm long, apically with a purple-red capitulum, pale green beneath. Inner corona are tassel-like; lobes purple-red , with a ring of nectar gland beneath. Disk is 1–2 mm high. Androgynophore is 8–10 mm long; stamens 5, flat; filaments free, flattened, about 1 cm long. Anthers are oblong, about 1.3 cm long. Ovary is ovoid. Styles are 3, free, purple-red, 1.6 cm long. Stigma is kidney-shaped. Berries are ovoid to subglobose, about 6 cm long, orange or yellow when ripe. Seeds are numerous, obcordate, about 5 mm long. Flowering is from May to July.

Distributed in: Pohnpei, Yap

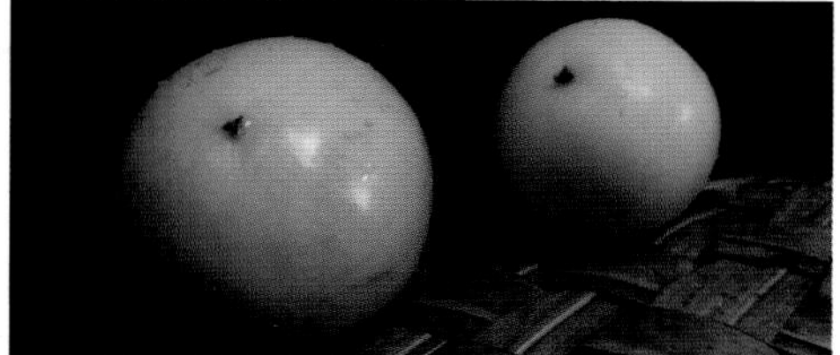

Football Fruit

Latin Name: ***Pangium edule*** **Reinw. ex. Bl.**

English Name: Pangi fruit, Football fruit

Yap Name: Rowal

Pohnpei Name: Durien

Moderate to large tree. Leaves are simple, entire, or 3 or many lobed, alternate, clustered at the end of twigs, cordate to ovate, 15–30 cm long. Petioles are long, rounded, subequal to leaves. Flowers are unisexual, about 5 cm wide; male flowers fascicled, female flowers solitary. Calyx lobes are 2 to 3; petals 5 to 6. Stamens are numerous in male flowers, 5 to 6 degerated (infertile) in female flowers; degenerated stamens alternate to petals; stigma sessile. Fruit is large, ovoid, 15 to 30 cm long, about half the width of the fruit. Rind is rough, brown; flesh yellow when ripe, musky, but delicious. Seeds are numerous, flat, about 5 cm long.

Distributed in: Pohnpei, Yap, Chuuk, Kosrae

Jackfruit

Latin Name: ***Artocarpus integra*** **Merr.**

English Name: Jackfruit

Evergree tree, 10–20 tall, 30–50 cm in breast height diameter. Old trees often have buttress roots. Bark is thick and dark brown. Twigs are 2–6 mm thick, wrinkled longitudinally to smooth, glabrous. Stipules are circular amplexcaul, with obvious traces. Leaves are leathery, spirally arranged, elliptic or obovate, 7–15 cm long or longer, 3–7 cm wide, obtuse or acuminate at apex, wedge-shaped at base, entire at margin when mature, or often lobed in young trees and sprouting shoots, dark green adaxially, pale green or pale brown when dry, glabrous, shiny, abaxially pale green and slightly rough, long-armed in mesophyll cells, with spherical or elliptic resin cells in tissue; lateral veins, pinnate, 6–8 on each side of the midrib; midrib prominently protruded abaxially. Petioles are 1–3 cm long. Stipules are oval, implexcaul, 1.5–8 cm long, appressed pilose or glabrous outside, deciduous. Plants are monoecious. Inflorescences are borne on old stems or short branches. Male inflorescences sometimes are axillary at the end of twigs or on short branches, cylindrical or clavate elliptic, 2–7 cm long; flowers numerous, some infertile ; peduncles 10–50 mm long; perianth of male flowers tubular, 1–1.5 mm long, 2 lobed in the upper part, puberulent; stamen 1; filaments erect in buds; anthers elliptic; pistillode absent. Female perianth is also tubular, toothed at apex, with base included in fleshy globose rachis; ovary 1 loculed. Syncarp is elliptic to spherical, or irregularly shaped, 30–100 cm long and 25–50 cm in diameter, pale yellow when young, yellowish brown when mature, with hard hexagonal tubercles and thick hairs on surface. Stones are oblong, about 3 cm in length and 1.5–2 cm in diameter. Flowering is from February to March.

Distributed in: Pohnpei

Pitaya

Latin Name: ***Hylocereus undulatus*** **Britt.**

English Name: Pitaya

Pohnpei Name: Dragon fruit

Perennial climbing succulent plants. Tap root is absent; lateral roots are widely distributed in the shallow soil layer; aerial roots numerous, grow by climbing. Stems are dark green, stout, up to 7 m long, 10–12 cm thick, 3 ribbed; Ribs flat, undulate in margin. Aerial roots are borne on nodes, can climb onto other plants for growth, mostly 3 ribbed, spinulose at impressed node. Leaves are degeranated due to long evolution in the tropical desert areas, whose photosynthesis function is assumed by the stem. The stem has a large number of parenchyma cells full of viscous liquid in the inner side, which are conducive to absorbing as much water as possible in the rainy season. Flowers are white, about 30 cm long; ovary large, inferior. Calyx is tubular, about 3 cm wide; lobes green, sometimes pale purple; scales 3–8 cm long. Petals are broad, pure white, erect, oblanceolate, entire. Stamens are numerous slender, up to 700–960, equal to or shorter than style. Anthers are creamy; filaments white; styles thick, 0.7–0.8 cm in diameter, creamy. Stigmas are 24 lobed. Fruits are oblong or oval, 10–12 cm long, fleshy; scales oval, acute at apex; skin red, thick, waxy; flesh white or red. Fruit setting is in summer and autumn.

Distributed in: Pohnpei, Yap, Chuuk, Kosrae

Fig

Latin Name: ***Ficus carica*** **Linn.**

English Name: Fig

Deciduous shrub, dioecious, 3–10 m tall, many branched. Barks are grayish brown with obvious lenticels. Branchlets are erect and stout. Leaves are alternate; leaf blade thickly papery, broadly ovate, subequal in length and width, 10–20 cm, usually 3-5-lobed; small lobes ovate, irregularly crenate at margin, adaxially scabrous, abaxially densely covered with small cystoliths and gray pubescence; shallow cordateat base; basal lateral veins 3–5; secondary veins 5–7 on each side of the midrib. Petiole is 2–5 cm long , stout. Stipules are ovate-lanceolate, 1 cm long, red. Male and gall flowers are symbiotic on the inner wall of a fruit. Male flowers borne on inner wall near apical pore; calyx lobes 4–5; stamens 3, sometimes 1 or 5. Gall flowers are short; style lateral. Female flowers are similar in calyx to male flowers; ovary smooth; style lateral; stigma 2-lobed, linear. Figs are axillary, solitary, large, reniform, 3–5 cm in diameter, apically depressed, purple-red or yellow when mature. Basal bracts are 3. Achenes are lens-like. Flowering and fruiting occur from May to July.

Distributed in: Pohnpei, Yap, Chuuk, Kosrae

Avocado

Latin Name: ***Persea americana*** **Mill.**

English Name: Avocado, Alligator per

Tree, about 10 m tall. Bark is grayish green, longitudinally fissured. Leaves are alternate; leaf blades oblong, elliptic, oval or obovate, 8–20 cm long and 5–12 cm wide, acute at apex, cuneate, acute to subcircular at base, leathery, adaxially green, usually abaxially slightly pale white, when young adaxially sparsely and abaxially densely yellowish-brown pubescent, when old adaxially glabrousand abaxially sparsely pubescent. Veins are pinnate; midribs adaxially depressed in the low part, flat in the upper part, abaxially prominently protruded; lateral veins 5–7 on each side of the midrib, slightly raised adaxially, very prominently protrudent abaxially; transverse veins and veinlets adaxially conspicuous, abaxially prominently protrudent. Petiole 2–5 cm long, slightly sulcate ventrally, slightly pubescent. Cymose panicles are 8–14 cm long, mostly inserted on the lower part of the branchlets, pedunculate; peduncles 4.5–7 cm long; peduncle and rachis densely yellowish brown pubescent; bracts and bracteoles filiform, about 2 mm long, densely yellowish brown pubescent. Flowers are pale green tinged with yellow, 5–6 mm long; pedicelsupto 6 mm long, densely yellowish brown pubescent. Perianth is densely yellow-brown pubescent on both surfaces; perianth tube obconical, about 1 mm long; perianth lobes 6, oblong, 4–5 mm long, obtuse at apex; outer 3 lobes smaller, dilated after anthesis, caducous. Fertile stamens are 9, 4 mm long; filaments filiform, complanate, densely pubescent, with no gland on the first and second whorls, with 1 pair of complanate, ovate and orange glands at base on the third whorl. Anthers are oblong, obtuse at apex, 4 loculed; locules introrse in the first and second whorls, extrorse in the third whorl. at bases staminodes are 3, located in the innermost whorl, sagittate cordate, about 0.6 mm long, glabrous, pedicelate; pedicels1.4 mm long, sparsely pubescent. Ovary is ovoid, 1.5 mm long, densely pubescent. Style is 2.5 mm long, densely pubescent. Stigma is slightly dilated, discoid. Fruits are large, usually pear-shaped, sometimes ovate or globose, sometimes oval or globose, 8–18 cm long, yellowish-green or reddish-brown. Exocarp iscorky; mesocarp fleshy and edible. Flowering is from February to March and fruiting from August to September.

Distributed in: Chuuk

Grape

Latin Name: ***Vitis vinifera*** **L.**

English Name: Grape

Woody vines. Branchlets are terete, longitudinally ribbed, glabrous or sparsely pubescent. Tendrils are bifurcated, opposite to each of two leaves. Leaves are ovate, conspicuously 3-5-lobed or moderately divided, 7–18 cm long and 6–16 cm wide, acute at apex of middle lobes; lobes are often close to each other, often constricted at base, with narrow or sometimes wide notch, deeply cordate at base, rounded at basal sinus, often connivent on both sides. Margins are 22–27 serrateded; teeth deep, thick, irregular, acute at apex. Lobes areadaxially green, abaxially palegreen, glabrous or sparsely pubescent. Basal veins are 5; lateral veins 4–5 on each side of the midrib; veinlets inconspicuously raised. Petiole is 4–9 cm long, almost glabrous; stipules caducous. Panicles are dense or sparse, many flowered, opposite to leaves, well developed at basal branches, 10–20 cm long; peduncle 2–4 cm long, almost glabrous or sparsely arachnoidly tomentous. Pedicles are 1.5–2.5 mm long, glabrous. Floral buds are obovate, 2–3 mm high, subrounded at apex. Calyx is discoid, undulate at margins, glabrous outside. Petals are 5, deciduous in a cap-shaped attachment. Stamens 5, filamentous, 0.6–1 mm long. Anthers yellow, oval, 0.4–0.8 mm long, significantly short and abortive or completely degenerated in female flowers. Disk is developed, 5 lobed. Pistil is 1, completely degraded in male flowers. Ovary is ovate; stigma short, enlarged. Fruit is spherical or elliptic, 1.5–2 cm in diameter. Seeds are obovate-elliptic, short subrounded at apex, shortly beaked at base. Chalaza is elliptical on the dorsal middle of seed; raphe slightly prominent, ventrally raised at raphal ridge, widely sulcate on ventral holes towards 1/4 of the seed. Flowering is from April to May and fruiting from August to September.

Distributed in: Pohnpei

Cashew

Latin Name: ***Anacardium occidentale*** **L.**

English Name: Cashew

Tree, 4–15 m tall. Branchlets are yellowish brown, glabrous or subglabrous. Leaves are leathery, obovate, glabrous on both surfaces, 8–14 cm long and 6–8.5 cm wide; apex is rounded, truncated or slightly concave; basebroadly wedge-shaped; margin entire. Lateral veins are 12 on each side of the midrib; lateral veins and reticulate veins protrudent on both surfaces. Petiole is 1–1.5 cm long. Panicles are broad, multi-branched, corymbose,10–20 cm long, many flowered, densely rusty sericeous. Bracts are ovate-lanceolate, 5–10 mm long, abaxially rusty minutely pubescent. Flowers are yellow, polygamous; pedicels absent orshort. Calyx is densely rusty sericeous outside; lobes ovate-lanceolate, acute at apex, 4 mm long, 1.5 mm wide. Petals are linear lanceolate, 7–9 mm long, about 1.2 mm wide, rusty sericeous outside, sparsely sericeous orglabrous inside, curled outward at anthesis. Stamens are 7–10, usually only 1 fertile, 8–9 mm long, 5–6 mm long in the perfect flowers. Stamenoids are shorter (3–4 mm long); filaments more or less connate at base; anthers small and oval; ovary obovate, 2 mm long, glabrous; style subulate, 4–5 mm long. Cashew nut is kidney-shaped, 2–2.5 cm long, 1.5 cm wide, attached on fleshy pear-shaped or turbinate pseudofruit at base. Pseudofruit is 3–7 cm long, 4–5 cm wide at the widest part, red or yellow when ripe. Kernel is kidney-shaped, 1.5–2 cm long and 1 cm wide.

Distributed in: Yap

Mulberry

Latin Name: ***Morus alba*** **Linn. var.** ***alba***

English Name: Mulberry

Tree or shrubs, 3–10 m tall or taller. Stem is up to 50 cm in breast height diameter. Bark is thick, gray, irregularly shallowly longitudinally fissures. Buds in winter are ovate, bud scales reddish-brown, imbricate, gray-brown, puberulent. Branchlets are puberulent. Leaves are ovate or broadly ovate, 5–15 cm long and 5–12 cm wide, acute, acuminate or obtuse at apex, rounded to shallowly cordate at base, coarsely serrate to crenate at margin, sometimes irregularly lobed, adaxially bright green and glabrous abaxially sparsely pubescent along main veins or in tufts in axils of the veins. Petiole is 1.5–5.5 cm long, pubescent. Stipules are lanceolate, caducous, densely hispidulous outside. Flowers are unisexual, axillary or axillary in bud scales, emerge concurrently with leaves. Male catkins are pendulous, 2–3.5 cm long, densely white pubescent, with male flowers; calyx lobes are broadly elliptic, pale green; filaments inflexed in bud; anthers 2-loculed, globose to reniform, longitudinally dehiscent. Female catkins are 1–2 cm long, pubescent; peduncles 5–10 mm long, pubescent; female flowers sessile; calyx lobes obovate, obtuse at apex, pubescent on the outside and margin, subtending ovary from both sides; style absent; stigma 2-lobed, papilose inside. Syncarp is ovate-elliptic, 1–2.5 cm long, red or dark purple at maturity. Flowering is from April to May and fruiting from May to August.

Distributed in: Pohnpei

Wampee

Latin Name: ***Clausena lansium*** **(Lour.) Skeels**

English Name: Wampee

Small tree, up to 12 m tall. Branchlets, leaf axes, rachis, especially the abaxial veins of unfolded leaflets, are scattered with many conspicuous oily fine protuberance and densely covered with short straight hairs. Leaves are 5-11- foliolate; leaflets ovate or oval-shaped, often oblique on one side, 6–14 cm long and 3–6 cm wide; base is nearly rounded or broadly wedge-shaped, asymmetrical on both sides; margins undulate or shallowly crenate; midribs usually adaxially pubescent. Petiole is 4–8 mm long. Inflorescences are terminal, paniculate; flower buds globose, with 5 slightly raised longitudinal ridges. Calyx lobes are ovate, about 1 mm long, pubescent outside. Petals are oblong, about 5 mm long, puberulent on both surfaces or glabrous adaxially. Stamens are 10, alternate in long and short stamens; long stamens equal in length to petals. Filaments are linear, slightly expanded at base, not geniculate. Ovary is densely hirsute, short-stalked; disk small. Fruit is rounded, ellipsoid, or broadly ovate, 1.5–3 cm long, 1–2 cm wide, pale yellow to dark yellow, hairy. Flesh is milky white, translucent. Seeds are 1–4; cotyledons dark green. Flowering is from April to May and fruiting from July to August.

Distributed in: Pohnpei, Yap

Canistel

Latin Name: ***Pouteria campechiana*** **(Kunth) Baehni,** ***Lucuma nervosa*** **A.DC**

English Name: Canistel, Egg fruit, Egg yolk fruit

Small tree, 6 m tall. Branchlets are terete, grayish-brown; twigs brown puberulent. Leaves are tough, papery, narrowly ellipsoid, 10–15 (20) cm long and 2.5–3.5 (4.5) cm wide, acuminate at apex, cuneate at base, glabrous on both surfaces. Midrib is slightly convex adaxially, rounded and very convex abaxially; lateral veins 13–16 on each side of the midrib, obliquely ascendant to the leaf edge and then articulately ascendant, conspicuous on both surfaces; tertiary veins reticulate, conspicuous on both surfaces. Petiole is 1–2 cm long. Inflorescence is 1 (2) flowered, axillary; pedicel terete, 1.2–1.7 cm long, covered by brown fine wooly hairs. Calyx lobes are usually 5, rarely 6–7, ovate or broadly ovate, about 7 mm long and 5 mm wide, slightly longer in the inner side, covered with creamy fine wooly hairs in the outer side, glabrous in the outer side. Corolla is longer than calyx, about 1 cm long, covered with yellow-white fine wooly hairs in the outer side, glabrous in the inner side. Corola tubes are cylindrical, about 5 mm long; corolla lobes (4)6, narrowly ovate, about 5 mm long. Stamens are usually 5 fertile; filaments subulate, about 2 mm long, covered with white fine wooly hairs. Anthers are cordate oval, about 1.5 mm long. Staminodes are narrowly lanceolate to subulate, 3 mm long, covered with white fine wooly hairs. Ovary is conical, 3–4 mm long, covered with yellowish-brown fine wooly hairs, 5 loculed; style cylindrical, 4–5 mm long, glabrous; stigma capitate. Fruit is obovate, about 8 cm long, green to yolk yellow, glabrous; pericarp very thin; mesocarp fleshy, thick, yolk yellow, edible, tastes like yolk and is thus called egg fruit or egg yolk fruit. Seeds are 2–4, elliptic, flattened, 4–5 cm long, yellowish brown, glossy; scars lateral, oblong, almost equal to seeds. Flowering is in spring and fruiting in autumn.

Distributed in: Pohnpei

Panama Cherry

Latin Name: ***Muntingia colabura*** **L.**

English Name: Panama cherry

Yap Name: Budo

Evergreen tree. Canopy is umbrella-shaped or vase-shaped. Branches are scattered. Barks are longitudinally fissured; lenticels conspicuous in old branches. Leaves are simple, alternate, papery, oblong, acute at apex, serrate at margin. Flowers are axillary; corolla white, usually 1–2 flowered. Fruits are berries, rounded, red to dark red when ripe, edible, sweet. Seeds are small.

Distributed in: Yap

Indian Jujube

Latin Name: ***Ziziphus mauritiana*** **Lam.**

English Name: Ber, Indian Jujube

Evergreen tree or shrub, up to 15 m tall. Young branches are densely yellow-gray tomentose; twigs pubescent; old branches purple-red. Stipular spines are 2, one oblique upward and the other hooklike recurved. Leaf blades are papery or thickly papery, oval, rectangular, elliptic, rarely subrounded, 2.5–6 cm long and 1.5–4.5 cm wide; apex round, rarely acute; base subrounded, slightly oblique, irregular; margin serrulate; adaxial surface dark green, glabrous, shiny; abaxial surface yellow or grayish-white tomentose; veins 3, basal, adaxially impressed or slightly prominent, abaxially conspicuously reticulate. Petiole is 5–13 mm long, densely grayish-yellow tomentose. Flowers are greenish yellow, bisexual, 5-petalled. Cymes are bifurcate, axillary, several or tens flowered, nearly not pedunculate or short pedunculate. Pedicels are 2–4 mm long, gray-yellow tomentose. Sepals are ovate-triangular, acute at apex, adaxially pubescent. Petals are oblong spathulate, claw-shaped at base. Stamens are subequal to petals; disk thick, fleshy, 10 lobed, centrally depressed; ovary globose, glabrous, 2 loculed, with one ovule each locule; Styles 2, shallowly lobed or half divided. Drupe is oblong rounded or globose, 1–1.2 cm long, about 1 cm in diameter, orange or red, black when ripe, with persistent calyx tube at base. Pedicel is 5–8 mm long, pubescent. Mesocarp is thin and corky; endocarp thick, hard-leathery. Seeds are 1 or 2, broad, flat, 6–7 mm long, 5–6 mm wide, reddish-brown, shiny. Flowering is from August to November and fruiting from September to December.

Distributed in: Pohnpei

Loquat

Latin Name: ***Eriobotrya japonica*** **(Thunb.) Lindl.**

English Name: Loquat

Evergreen and small tree, up to 15 m tall. Branchlets are stout, yellowish brown, densely rusty or gray brown tomentose. Leaf blade is leathery, lanceolate, oblanceolate, obovate or elliptic-oblong, 12–30 cm long, 3–9 cm wide; apex acute or acuminate; base cuneate or tapered to form petiole; margin sparsely serrated in the upper part, entire at base; upper surface glossy, rugose; lower surface densely grey brown tomentose; lateral veins 11–21 on each side of the midrib. Petiole is short or almost sessile, 6–10 mm long, gray-brown tomentose. Stipules are 1–1.5 cm long, acute at apex, pubescent. Panicles are terminal, 10–19 cm long, many flowered. Peduncles and pedicels are densely rusty tomentose; pedicel 2–8 mm long. Bracts are 2–5 mm long, subulate, densely rusty tomentose. Flowers are 12–20 mm in diameter. Calyx tube is shallowly cupular, 4–5 mm long; calyx lobes triangular-ovate, 2–3 mm long, acute at apex; calyx tubes and lobes adaxially rusty tomentose. Petals are white, oblong or ovate, 5–9 mm long, 4–6 mm wide, clawed at base, rusty tomentose. Stamens are 20, much shorter than petals; filaments expanded at base. Styles are 5, free; stigma capitate, glabrous; ovary rusty pubescent at apex, 5 loculed with 2 ovules each locule. Fruits are spherical or oblong, 2–5 cm in diameter, yellow or orange, adaxially rusty pubescent, fall off soon. Seeds are 1–5, spherical or oblate, 1–1.5 cm in diameter, brown, glossy; seed coat papery. Flowering is from October to December and fruiting from May to June.

Distributed in: Pohnpei

Watermelon

Latin Name: ***Citrullus lanatus*** **(Thunb.) Matsum. & Nakai,** ***Citrullus colocynthis*** **(L.) Kunt**

English Name: Watermelon

Pohnpei Name: Sika

An annual vine. Stems and branches are stout, conspicuously angled, dense white or yellowish-brown pilose. Tendrils are thick, pubescent, bifurcate. Petioles are thick, 3–12 cm long, 0.2–0.4 cm wide, inconspicuously furrowed, densely pubescent. Leaf blades are papery, triangular ovate, white-green, 8–20 cm long, 5–15 cm wide, shortly hispid on both surfaces especially on veins and on lower surface, 3 partite; midlobe longer, obovate, oblong lanceolate or lanceolate, acute or acuminate at apex; lobes pinnate or bipinnate, shallowly or deeply lobed, wavy or sparsely toothed, usually shallowly a few serrated on last lobe; apex obtuse; base cordate, sometimes semicircularly sinuate; sinus 1–2 cm long and 0.5–0.8 cm deep. This vine is monoecious. Both male and female flowers are solitary, axillary; for male flowers, pedicels are 3–4 cm long, densely yellowish-brown pilose; calyx tube broadly bell-shaped, densely pilose; calyx lobes are narrowly lanceolate, subequal

to the length of calyx tube, 2–3 mm long; corolla pale yellow, 2.5–3 cm in diameter, tinged with green on outer surface, pilose; lobes ovate-oblong, 1–1.5 cm long, 0.5–0.8 cm wide; apex obtuse or slightly pointed; veins yellowish-brown, pubescent; stamens 3, nearly free, filaments short; anther cell reflexed. For female flowers calyx and corolla are the same as those of male flowers; ovary is ovate, 0.5–0.8 cm long, 0.4 cm wide, densely pilose; style 4–5 mm long; stigma 3, reniform. Fruit is large, nearly spherical or elliptical, fleshy, juicy; rind smooth, various in colors and mottling or striping. Seeds are numerous, ovate, black or red, sometimes white, yellow, light green or striped, smooth on both surfaces, obtuse atbase, usually slightly arched on edge, 1–1.5 cm long, 0.5–0.8 cm wide, 1–2 mm thick. Flowering and fruiting stages are in summer.

Distributed in: Pohnpei, Yap, Chuuk, Kosrae

Winter Cassaba Melon

Latin Name: ***Cucumis melo*** **L. var.** ***inodorus*** **Naud.**

English Name: Winter cassaba melon

Fruit is oblong cylindrical or nearly clavate, 20–30 (50) cm long and 6–10 (15) cm in diameter, slightly thicker on the upper part than on the lower part, round or slightly truncated at both ends, smooth, glabrous, pale green, vertically lined; flesh white or pale green, not sweet. Flowering and fruiting occur in summer.

Distributed in: Pohnpei, Yap

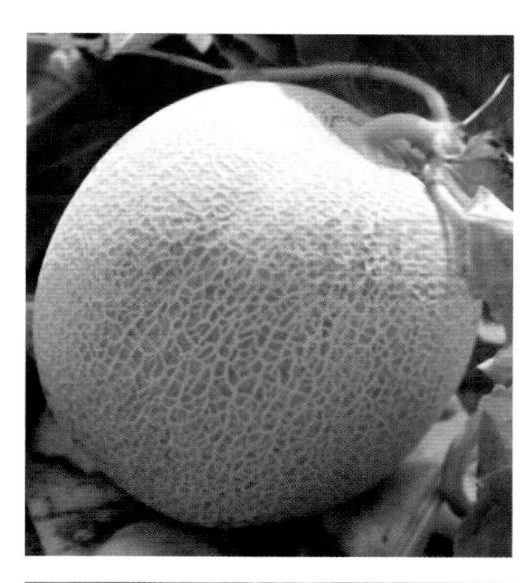

Muskmelon

Latin Name: *Cucumis melo* L. var. *melo*

English Name: Cantaloupe, Muskmelon

Annual creeping or climbing herb. Stem and branches are angled, yellowish-brown or white hispid, scabrous. Tendrils are slender, single, and puberulent. Petiole is 8–12 cm long, grooved, short setose. Leaf blades are thick papery, subcircular or kidney-shaped, 8–15 cm long and wide, adaxially coarse and white hispid, abaxially densely hispid along veins, undivided or 3-7-lobed, palmately veined; lobes obtuse at apex, serrated, truncate or semicircularly sinuous at base. Plants are monoecious; flowers are unisexual. Male flowers are axilliary, fascicled; pedicels slender, 0.5–2 cm long, pubescent; calyx tube narrowly bell-shaped, densely white villous, 6–8 mm long; calyx tube subulate, erect or spread, shorter than calyx tube; corolla yellow, 2 cm long; corolla lobes ovate-oblong, acute at apex; stamens 3; filaments very short; anther locule flexuous; connetives elongated at apex. Staminodes are about 1 mm long. Female flowers are solitary; pedicles coarse, pubescent. Ovary is oblong elliptic, densely pilose and hispid; styles 1–2 mm long; stigma about 2 mm. Fruits vary in shape and color dependent on varieties, are usually spherical or elongate elliptic; rind smooth, with longitudinal grooves or stripes, without spiny protuberances; flesh white, yellow or green, sweet. Seeds are dirty white or yellow white, oval or oblong, acute at apex, obtuse at base, smooth on surface, immarginate. Flowering and fruiting occur in summer.

Distributed in: Yap

Vegetables

Cucumber

Latin Name: ***Cucumis sativus*** **L.**

English Name: Cucumber

Annual vine or climbing herb. Stems and branches are elongated, furrowed, white hispid. Tendrils are slender, simple, white pubescent. Petioles are scabrid, hispid, 10–16 (20) cm long. Leaf blades are broadly ovate cordate, membranous, 7–20 cm in length and width, scabrous on both surfaces, hispid, 3–5 angled or shallowly lobed; lobes triangular, dentate; margins sometimes ciliate; apex acute or acuminate; basal sinus semicircular, 2–3 cm wide, 2–2.5 cm deep, sometimes dorsally contiguous. Plants are monoecious. Male flowers are axillary, usually fascicuate with a few flowers; pedicels slender, 0.5–1.5 cm long, puberulent; calyx tube narrowly campanulate or nearly cylindrical, 8–10 mm long, densely white pilose; calyx lobes subulate, spreading, subequal to calyx tube; corolla yellow-white, about 2 cm long; corolla lobes oblong lanceolate, acute at apex. Stamens are 3; filaments almost absent; anthers 3–4 mm long; connectives extended, about 1 mm long. Female flowers are solitary or sparsely fascicled; pedicels stout, pubescent, 1–2 cm long. Ovary is spindle-shaped, muricate, spinulose. Fruits are oblong or cylindrical, 10–30 (50) cm long. Yellowish-green when ripe, muricate, spinulose, very rarely nearly smooth. Seeds are small, narrowly ovate, white, immarginate, nearly acute at both ends, about 5–10 mm long. Flowering and fruiting occur in summer.

Distributed in: Pohnpei, Yap, Chuuk, Kosrae

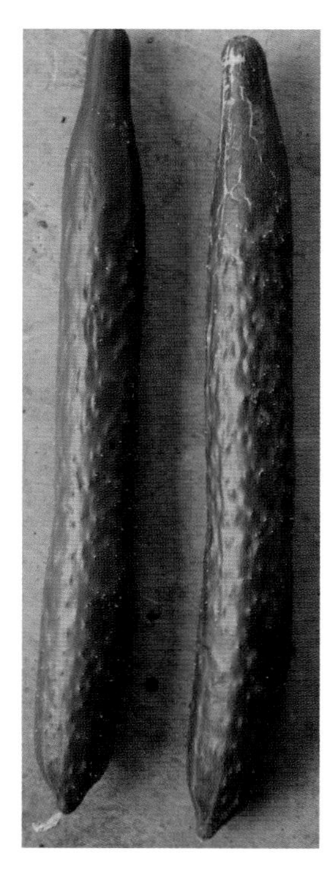

Bitter Gourd

Latin Name: ***Momordica charantia*** **L.**

English Name: Bitter gourd

Plants are annual scandant, many branched. Stems and branches are pubescent. Tendrils are filiform, 20 cm long, puberulent, not branched. Petiole is slender, initially white pubescent, nearly glabrous at a late stage, 4–6 cm long. Leaf blades are is ovate reniform or suborbicular, membranous, 4–12 cm in length and width, adaxially green, abaxially pale green, densely conspicuously puberulent on veins, sparsely puberulent, 5–7 partite; lobes ovate-oblong; margins crenate or irregularly lobed; apex mostly obtuse, rarely acute; basal sinus, semicircular; veins palmate. Plants are monoecious. Male flowers are solitary, axillary, puberulent; pedicels slender, 3–7 cm long, with 1 bract on the middle or lower part. Bracts are green, kidney-shaped or round, entire, slightly ciliated sparsely pubescent on both surfaces, 5–15 mm in length and width. Calyx lobes are ovate-lanceolate, white pubescent, 4–6 mm long, 2–3 mm wide, acute at apex. Corolla is yellow; lobes obovate, obtuse, acute or retuse at apex, 1.5–2 cm long, 0.8–1.2 cm wide. Stamens are 3, free; anther locules 2 conduplicate. Female flowers are solitary; pedicel puberulent, 10–12 cm long, with 1 bract at base; ovary spindle-shaped, densely verrucous; stigma 3, expanded, 2 lobed. Fruit is spindle-shaped or cylindrical, verrucose, 10–20 cm long, orange-yellow when ripe, 3 valved at apex. Seeds are numerous, oblong, with red aril, 3 small toothed at each end, incised on both surface, 1.5–2 cm long, 1–1.5 cm wide. Flowering and fruiting occur from May to October.

Distributed in: Pohnpei, Yap, Chuuk, Kosrae

Wax Gourd

Latin Name: ***Benincasa hispida*** **(Thunb.) Cogn.var.** ***hispida***

English Name: Wax gourd

Annual climbing or trailing vine. Stem is yellowish-brown hispid and pilose, furrowed. Petiole is stout, 5–20 cm long, yellowish-brown hispid, and pilose. Leaf blades are reniformed, suborbicular, 15–30 cm wide, 5–7 shallowly lobed or sometimes moderately lobed; lobes broadly triangular or ovate, acute at apex, dentate at margin, deeply cordate at base; sinus open, nearly circular, 2.5–3.5 cm depth and width; adaxial surface dark green, scabrid, sparsely pubescent, caduceus and glabrescent after aging; abaxial surface scabrous, gray-white, with coarse bristles. Veins are slightly raised on the lower surface densely hairy. Tendrils have 2–3 furcate, hispid, pilose. Plants are mon-oecious. Flowers are solitary. Male pedicels are 5–15 cm long, densely yellowish-brown hispid and pilose, often with a bract at base; bracts oval or broadly oblong, 6–10 mm long, acute at apex, pilose. Calyx tube is broadly bell-shaped, 12–15 mm wide, densely setose and pilose; calyx lobes lanceolate, 8–12 mm long, serrated, reflexed. Corolla is yellow, rotate; lobes broadly obovate, 3–6 cm long, 2.5–3.5 cm wide, sparsely pubescent on both surfaces, obtuse at apex, 5 veined. Stamens are 3, free; filaments 2–3 mm long, swollen at base, haired. Anthers are 5 mm long, 7–10 mm wide; anther cells are 3 conduplicate. Female pedicels are shorter than 5 cm, densely yellowish-brown hispid and pilose. Ovary is oval or cylindrical, densely covered with yellowish-brown wooly brittles, 2–4 cm long. Style is 2–3 mm long; stigma 3, 12–15 mm long, 2 lobed. Fruit is large, long cylindrical or subglobose, hirsute, white pruinous, 25–60 cm long, 10–25 cm in diameter. Seeds are ovate, white or pale yellow, flat, margined, 10–11 mm long, 5–7 mm wide, 2 mm thick.

Distributed in: Pohnpei, Yap, Chuuk, Kosrae

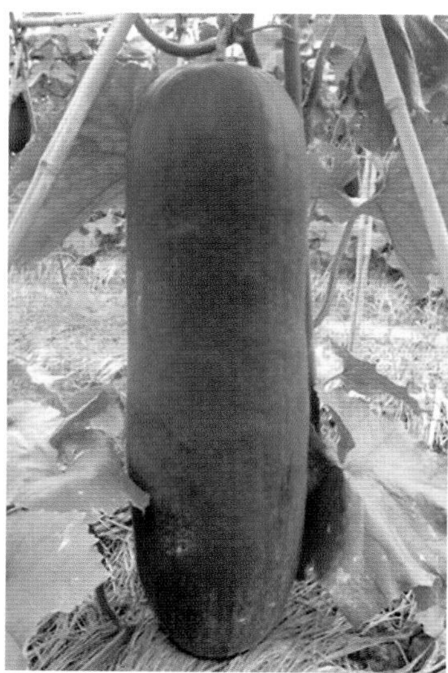

Ash Gourd

Latin Name: ***Benincasa hispida*** **(Thunb.) Cogn. var.** ***chieh-qua*** **How**

English Name: Ash Gourd

The difference from wax gourd (the original variety) lies in that living ovary is dirty-colored or yellow hispid, that fruits are smaller, slightly longer and thicker than cucumber, 15–20 (25) cm long and 4–8 (10) cm in diameter, hispid when mature, not coated with white waxy powder.

Distributed in: Pohnpei, Yap, Chuuk, Kosrae

Smooth Luffa

Latin Name: *Luffa cylindrica* (L.) Roem.

English Name: Angled gourd, Sponge gourd

Annual climbing vines. Stems and branches are rough, furrowed, puberulent. Tendrils are slightly stout, pubescent, usually 2–4 furcate. Petiole is rough, 10–12 cm long, inconspicuously furrowed, subglabrous. Leaf blades are triangular or subrounded, 10–20 cm in length and width, usually 5–7 palmately lobed; lobes triangular; middle lobe longer, 8–12 cm long; apex acute or acuminate; margin serrated; basedeeply cordate; sinus 2–3 cm deep, 2–2.5 cm wide. Adaxial surface dark green, rough, scabrous; abaxial surface pale green, pubescent; veins palmate, white pubescent. Plants are monoecious. Male flowers are usually 15–20 flowered, borne on the upper part of the raceme; peduncle slightly stout, 12–14 cm long, pubescent; pedicel 1–2 cm long; calyx tubes broadly bell-shaped, 0.5–0.9 cm in diameter, pubescent; calyx lobes ovate-lanceolate or subtriangular, reflexed outward at the upper part, about 0.8–1.3 cm long, 0.4–0.7 cm wide, densely pubescent on the inner surface, especially at margin, less pubescent on the outer surface, acuminate at apex, 3 veined. Corolla is yellow, rotate, 5–9 cm in diameter when open; lobes oblong, 2–4 cm long, 2–2.8 cm wide, densely yellowish-white pilose at base on the inner part, 3–5 prominently veined on outer surface; veins densely pubescent, obtuse at apex and narrow at base. Stamens are usually 5, rarely 3. Filaments are 6–8 mm long, white pubescent at base, slightly connate at first and then free; anther cell is conduplicate. Female flowers are solitary; pedicel 2–10 cm long; ovary long cylindrical, pubescent; stigma 3, enlarged. Fruit is cylindrical, straight or slightly curved, 15–30 cm long and 5–8 cm in diameter, smooth, usually dark longitudinally striped, fleshy when immature, dry when mature, reticulately fibrous inside. Seeds are numerous, black, ovate, flat, smooth, narrowly winged on the edge. Flowering and fruiting are in summer and autumn.

Distributed in: Pohnpei, Yap, Chuuk, Kosrae

Chinese Pumpkin

Latin Name: ***Cucurbita moschata*** **(Duch. ex Lam.) Duch. ex Poiret**

English Name: Chinese pumpkin

An annual vine. Stems are usually nodular, 2–5 m when stretched, densely covered with white short setae. Petiole is stout, 8–19 cm long, short setose. Leaf blade is ovate or oval, slightly soft, 5-angled or 5-lobed, rarely obtuse, 12–25 cm long, 20–30 cm wide; lateral lobes small; middle lobe large; lobes triangular; adaxial surface densely covered with yellow-white bristles and velvets, usually with some white spots, prominently veined, with midrib of each lobe often extending to apexe to form a small point; abaxial surface pale, more obviously haired, covered with small and dense fine teeth on the edge; slightly obtuse at apex. Tendrils are slightly robust, short setose and tomentose like petioles, 3–5 furcate. Plants are monoecious. Male flowers are solitary; calyx tube bell-shaped, 5–6 mm long; calyx lobes pubescent, striped, 1–1.5 cm long, expanded into leaf shape at the upper part. Corolla is yellow, bell-shaped, 8 cm long, 6 cm in diameter, 5 lobed, revolute at margin, rugose, acute at apex. Stamens are 3; filaments glandular, 5–8 mm long. Anthers are connivent, 15 mm long, with tortuous cells. Female flowers are solitary; ovary 1 loculed; style short; stigma 3, enlarged, bifid at apex. Pedicel is stout, angled, sulcate, 5–7 cm long, enlarged into trumpet-shape at apex. Fruits are variable in shape, outside often longitudinally sulcate or not sulcate. Seeds are numerous, long oval or oblong, gray-white, thin at margin, 10–15 mm long, 7–10 mm wide.

Distributed in: Pohnpei, Yap, Chuuk, Kosrae

A33
A2

Tomato

Latin Name: ***Lycopersicon esculentum*** **Mill.**

English Name: Tomato

Annual herb, 0.6–2 m tall, viscid pubescent, odorous. Stems lodge easily. Leaves are pinnately compound or pinnately partite, 10–40 cm long; leaflets 5–9, irregular, variable in size, oval or rectangular, 5–7 cm long; margins irregularly serrated or lobed. Peduncle is 2–5 cm long, often 3–7 flowered; pedicel 1–1.5 cm long. Calyx is rotate, lanceolate, persistent when fruiting; corolla rotate, yellow, about 2 cm in diameter. Berries are oblate or subglobose, fleshy and juicy, orange or bright red, glossy. Seeds are yellow. Flowering and fruiting occur in winter and spring.

Distributed in: Pohnpei, Yap, Chuuk, Kosrae

Hot Pepper

Latin Name: ***Capsicum annuum*** **L.var.** ***annuum***

English Name: Hot pepper

Annual or perennial plant, 40–80 cm tall. Stems are glabcresent or puberulent, zigzag curved at twigs of lateral branches. Leaves are alternate, paired or fascicled, not elongated at apical node of the twig; leaf blade oblong ovate, ovate or ovate lanceolate, 4–13 cm long and 1.5–4 cm wide; margin entire; apex short acuminate or acute apex; base narrowly cuneate. Petiole is 4–7 cm long. Flowers are solitary and drooping. Calyx is cup-shaped, inconspicuously 5-toothed. Corolla is white; lobes ovate. Anthers are grayish purple. Pedicels are stout and drooping; fruits are long fingered; apex acuminate, often curved, green when immature, red, orange or purple when mature, spicy. Seeds are pale yellow, flat reniform, 3–5 mm long. Flowering and fruiting stages are from May to November.

Distributed in: Pohnpei, Yap, Chuuk, Kosrae

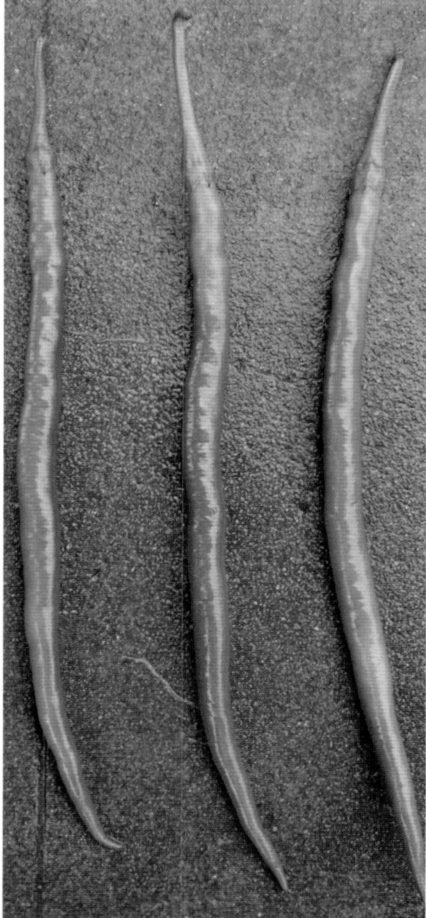

Sweet Pepper

Latin Name: ***Capsicum annum*** **L.var.** ***grossum*** **(L.)**

English Name: Sweet pepper

Plants are stout and tall. Leaf blades are oblong or ovate, 10–13 cm long. Pedicel is erect or cernuous. Fruits are large subglobular, cylindrical or oblate, longitudinally sulcate truncated or slightly depressed at apex, truncated and often slightly inwardly concave at base, not spicy but slightly sweet or slightly peppery.

Distributed in: Pohnpei, Yap, Chuuk, Kosrae

Cone Pepper

Latin Name: ***Capsicum annum*** **L. var.** ***conoides*** **(Mill.) Irish**

English Name: Cone pepper

For most of the plants branches are bifurcated. Leaf blades are 4–7 cm long, ovate. Flowers are often solitary between bifurcations, slightly cernuous. Pedicels are erect. Corolla is white or tinged with purple. Fruits are erect, small, conical, about 1.5 (3) cm long, red or purple after mature, very spicy.

Distributed in: Pohnpei, Yap, Chuuk, Kosrae

Asparagus lettuce

Latin Name: ***Lactuca sativa*** **L. var.** ***angustata*** **Irish ex Bremer**

English Name: Asparagus lettuce

Annual or biennial herb, 25–100 cm tall. Roots are vertically downward grown. Stems are erect, solitary, branched in panicle-like inflorescence at the upper part; all stem branches white. Basal and lower cauline leaves are large, not lobed, oblanceolate, elliptic or elliptic oblanceolate, 6–15 cm long, 1.5–6.5 cm wide, acute, short acuminate or circular at apex, cordate or sagittate semi-amplexicaul, undulate or serrated at margin; leaves in the upper part attenuate, similar in hape to the basal and lower cauline leaves, or lanceolate; leaves on lower branches of panicles and on branches of panicles very small, oval-cordate, sessile, cordate or sagittate amplexicaul, entire at margin, glabrous on both surfaces. Heads are numerous or most, arranged in panicles at the apical branches. Involucre is ovoid when fruiting, 1.1 cm long and 6 mm wide. Involucral bracts are of 5 layers; outermost layer triangular, about 1 mm long, 2 mm wide; the outer layer triangular or lanceolate, 5–7 mm long, 2 mm wide; the middle layer lanceolate to oval lanceolate, 9 mm long and 2–3 mm wide; the inner layer linear long elliptic, 1 cm long, about 2 mm wide. Involucral bracts are all acute at apex, glabrous outside. Florets are about 15, lingulate. Achenes are oblanceolate, 4 mm long and 1.3 mm wide, flattened, pale brown, 6–7 veined on each surface, acute at apex to form a fine beak. Beaks are filamentous, about 4 mm long, almost as long as achenes. The coronal hairs are 2 layered, slender, minutely hispid. Flowering and fruiting stages are from February to September.

Distributed in: Pohnpei, Yap, Chuuk, Kosrae

Lettuce

Latin Name: *Lactuca sativa* L. var. *ramosa* Hort.

English Name: Lettuce

Cabbage

Latin Name: ***Brassica oleracea*** **var.** ***capitata*** **L.**

English Name: Cabbage, Savoy

Biennial herb, pruinose. Annual stems are short, stout, fleshy, unbranched, green or grayish green; basal leaves numerous, thick, overlapping into a globose head, oblate, 10–30 cm in diameter or larger, milky white or pale green. Biennial stems are branched, cauline; basal or lower cauline leaves oblong-obovate to orbicular, 30 cm long and wide. Leaf apex is rounded; base abruptly narrowed into a very short broadly winged petiole; margin undulate, inconspicuously serrated. Upper cauline leaves are ovate or oblong-ovate, 8–13.5 cm long, 3.5–7 cm wide, amplexicaul; uppermost leaves oblong, about 4.5 cm long, about 1 cm wide, amplexicaul. Racemes are terminal, axillary. Flowers are pale yellow, 2–2.5 cm in diameter. Pedicels are 7–15 mm long. Sepals are erect, linear oblong, 5–7 mm long. Petals are broadly elliptic, obovate or suborbicular, 13–15 mm long, conspicuously veined, retuse at apex, suddenly narrowed into 5–7 mm long claws. Fruits are silique, cylindrical, 6–9 cm long, 4–5 mm wide, slightly flattened on both sides, prominent at midrib; beak conical, 6–10 mm long. Pedicels are thick, erectly spread, 2.5–3.5 cm long. Seeds are brown, globose, 1.5–2 mm in diameter. Flowering is in April and fruiting in May.

Distributed in: Pohnpei, Yap, Kosrae

Cauliflower

Latin Name: ***Brassica oleracea*** **var.** ***botrytis*** **L.**

English Name: Cauliflower

Biennial herb, 60–90 cm tall, pruinous. Stems are erect, stout, branched. Basal or lower cauline leaf blades are gray-green, oblong to elliptic, 2–3.5 cm long; apex rounded, spreading, entire or serrated, sometimes decurrent with several small lobes like wing. Petioles are 2–3 cm long. Leaf blades in the upper and middle of the stem are small, sessile, oblong to lanceolate, amplexicaul. Heads are fleshy, creamy white at apex, developed from compact inflorescence composed of peduncle, pedicels and undeveloped flower buds. Racemes are terminal, axillary. Flowers are pale yellow at the beginning and then white. Silique is cylindrical, 3–4 cm long, midveined, thick at upper beak, slender at lower beak, 10–12 mm long. Seeds are broadly elliptic, nearly 2 mm long, brown. Flowering is in April and fruiting in May.

Distributed in: Pohnpei, Yap

Chinese cabbage

Latin Name: ***Brassica campestris*** **L. ssp.** ***pekinensis*** **(Lour.) Olsson**

English Name: Chinese cabbage

Biennial herb. Plants are 40–60 cm tall, often glabrous, few setose on the abaxial leaf midrib. Basal leaves are numerous, large, obovate-oblong to broadly obovate, 30–60 cm long, as wide as less than half of the length; apex obtuse; margin crinkled, wavy, sometimes inconspicuously toothed; midrib white and very wide, with numerous stout lateral veins. Petiole is white, flattened, 5–9 cm long, 2–8 cm wide, with notched wide thin wings at margin. The upper cauline leaves are oblong-oval, oblong-lanceolate to long-lanceolate, 2.5–7 cm long, obtuse to short acute at apex, entire or dentate at margin, stalked or amplexicaul, pruinose. Flowers are bright yellow, 1.2–1.5 cm in diameter; pedicels 4–6 mm long; sepals erect, oblong or ovate–lanceolate, 4–5 mm long, pale green to yellow; petals obovate, 7–8 mm long, narrowed into claws at base. Fruits are silique, short, thick, 3–6 cm long, 3 mm wide, flattened on both sides; beak 4–10 mm long, 1 mm wide, rounded at apex; pedicel spreading or ascending, 2.5–3 cm long. Seeds are brown, globose, 1–1.5 mm in diameter. Flowering is May and fruiting in June.

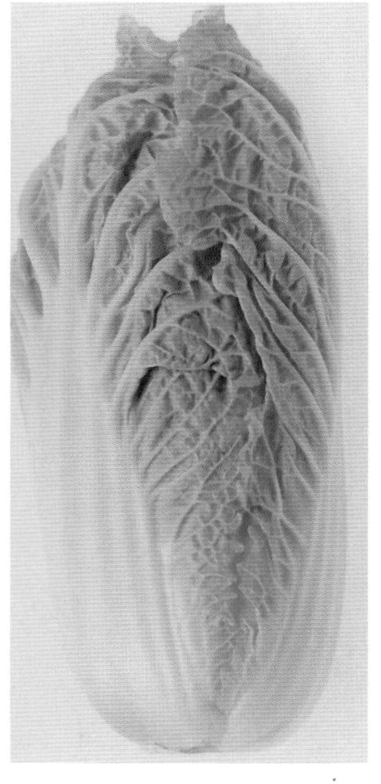

Distributed in: Pohnpei, Yap, Chuuk, Kosrae

Pak Choi

Latin Name: ***Brassica campestris*** **L. ssp.** ***chinensis*** **Makino**

English Name: Pak choi

Pak choi is originated from China and has been widely cultivated as a leafy vegetable, or sometimes escapes into the wild. It is commonly cultivated in the central and southern provinces of China while relatively rare in the north. It is suitable to grow in places with a warm and humid climate and a fertile humid soil. It is not cold-tolerant. Stems are cold injured, and leaves die at frost. It is distributed throughout tropical Asia, Africa and Oceania.

Distributed in: Pohnpei, Yap, Chuuk, Kosrae

Carrot

Latin Name: ***Daucus carota*** **L. var.** ***sativa*** **Hoffm.**

English Name: Carrot

Biennial herb, 15–120 cm tall. Stems are solitary, white hispid. Basal leaves are membranous, oblong, bipinnately or tripinnately divided; terminal segments linear or lanceolate, 2–15 mm long and 0.5–4 mm wide, acute at apex with a small tip, smooth or hispid. Petiole is 3–12 cm long. Cauline leaves are almost sessile, sheathed; ultimate segments small or slender. Inflorescence is compound umbel; peduncle 10–55 cm long, hispide. The involucre is composed of numerous bracts; bracts foliaceous, pinnately partite, rarely not partite; lobes are linear, 3–30 mm long; rays numerous, 2–7.5 cm long, curved inward at the outer part when fruiting. Bracteols are 5–7, linear, undivided or 2-3-lobed; margin membranous, ciliated. Flowers are usually white, sometimes tinged with reddish. Petioles are unequal, 3–10 mm long. Fruits are oval, 3–4 mm long, 2 mm wide, with white spiny hairs on edges. Roots are fleshy, long conical, thick, red or yellow. Flowering is from May to July.

Distributed in: Pohnpei, Yap, Chuuk, Kosrae

Chinese Chives

Latin Name: ***Allium tuberosum*** **Rottler**

English Name: Chinese chives

Herbaceous perennial plants. Rhizomes are obliquely creeping. Bulbs are clustered, subcylindrical; tunic dark yellow to yellowish brown, fibrous, reticulate or nearly reticulate. Leaves are striped, flat, solid, shorter than scape, 1.5–8 mm wide, smooth at margin. Scape is cylindrical, often 2-longitudinally ridged, 25–60 cm tall, sheathed at base. Involucral spathe is unilaterally valved, or 2- or 3-lobed, persistent. Umbel is hemispherical or subglobular, sparsely many flowered. Pedicles are subequal, 2–4 times longer than bracts, bracteolate at base, with several pedicels surrounded by a common bract at base. Flowers are white. Tepals often have green or yellowish-green midveins; inner tepals obovate-obovate, rarely ovate-obovate, mucronate or obtuse at apex, 4–7 (8) mm long, 2.1–3.5 mm wide; outer tepals usually narrow, oblong oval to oblong lanceolate, mucronate at apex, 4–7 (8) mm long, 1.8–3 mm wide. Filaments are equal in length, 2/3–4/5 of tepal length, connate at base, adnate to perianth segments, 0.5–1 mm tall at connate part, narrowly triangular at free part; slightly wider at base in the inner than in the outer ones. Ovary is obconical-globose, 3 round angled, minutely tuberculate on outer wall. Flowering and fruiting are in between July and September.

Distributed in: Pohnpei, Yap

Celery

Latin Name: ***Apium graveolens*** **L.**

English Name: Celery

Biennial or perennial herb, 15–150 cm tall, strongly aromatic. Roots are conical, brown, with numerous lateral roots. Stems are erect, smooth, a few branched, anguled, longitudinally grooved. Basal leaves are petiolate, 2–26 cm long, slightly enlarged into membranous sheaths at base. Leaf blade is oblong to obovate, 7–18 cm long, 3.5–8 cm wide, usually 3 partite or 3 divided; lobes nearly rhombic, cremate or serrated at margin, with veins raised on both surfaces. Upper cauline leaf blades shortly petiolate, broadly triangular, usually 3 parted; leaflets obovate, sparsely obtusely serrated to dissected above middle margin. Umbels are compound, terminal or opposite to leaves. Peduncles are variable in length, sometimes obsolete, usually bractless and ebracteolate; rays 3–16, 0.5–2.5 cm long, slender. Umbellules are7–29 flowered; pedicels 1–1.5 mm long. Calyx teeth are small or not obvious. Petals are white or yellowish green, ovate, 1 mm long, 0.8 mm wide, ligulately inflexed at apex. Filaments are equal to or slightly longer than petals; anthers ovate, about 0.4 mm long; style oblate at base, very short when young, about 0.2 mm long and recurved when mature. Schizocarp are rounded or oblong ellipsoid, 1.5 mm long, 1.5–2 mm wide, acute at edge, slightly constricted at connate surface, with 1 oil tube in each groove and 2 oil tubes in connate surface. Endosperm is ventrally flat. Flowering is from April to July.

Distributed in: Pohnpei, Yap

Eggplant

Latin Name: ***Solanum melongena*** **L.**

English Name: Garden eggplant, Eggplant

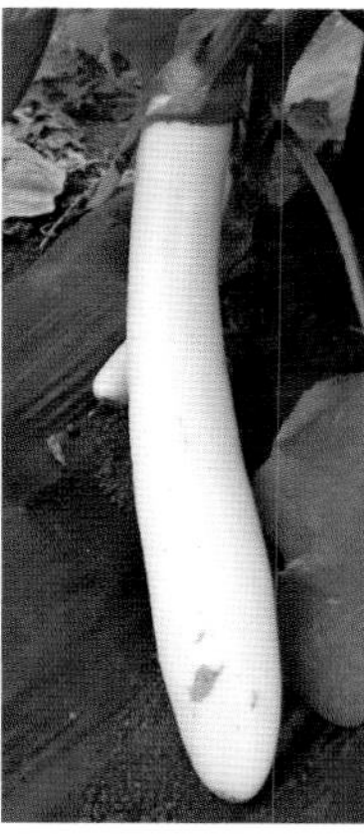
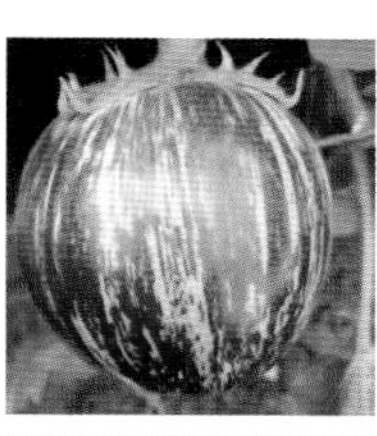

Erectly branched herbs to subshrubs, up to 1 m tall. Branchlets, petioles and pedicels are all 6–8 (10) branched, appressed or short-stalked stellate tomentose. Branchlets are mostly purple (wild ones often pricky), and tomenta gradually falls off as they grow older. Leaves are large, ovate to oblong-ovate, 8–18 cm long or longer, 5–11 cm wide or wider, obtuse at apex, unequal at base, repand or sinuate lobed at margin, adaxially densely covered with 3–7 (8) branched short and flat stellate tomenta, abaxially densely covered with 7–8 branched long and flat stellate tomenta. Lateral veins are 4–5 on each side of the midrib, adaxially sparsely and abaxially densely stellate tomentose; midrib is similar to the lateral veins in tomentum (both midvein and lateral veins of wild types have small prickles on both surfaces). Petiole is about 2–4.5 cm long (the wild one has prickles). Fertile flowers are solitary; pedicels about 1–1.8 cm long, densely tomentose, often pendulous after flowering. Sterile flowers are scorpioid, borned together with the fertile ones. Calyx is subcampanulate, 2.5 cm in diameter or slightly larger, densely tomentose and spinulose similar to pedicels; prickles about 3 mm long; calyx lobes lanceolate, acute at apex, sparsely stellate tomentose in the inner side. Corolla is rotate, densely stellate tomentose in the outer side, sparsely tomentose only at apex in the inner side; corolla tube about 2 mm long; limb about 2.1 cm long; corolla lobes triangular, about 1 cm long. Filament is about 2.5 mm long. Anther is about 7.5 mm. Ovary is round, densely stellate tomenose at apex. Style is 4–7 mm long, stellate tomentose below the middle; stigma shallowly lobed. Fruit varies greatly in shape and size.

Distributed in: Pohnpei, Yap, Chuuk, Kosrae

Flowering Chinese Cabbage

Latin Name:*Brassica parachinensis* Bailey

English Name: Flowering Chinese cabbage

Biennial herb, 30–90 cm tall. Stems are stout, erect, branched or unbranched, glabrous or subglabrous, slightly pruinose. Basal leaves are pinnately lobed with an enlarged terminal lobes; terminal lobes round or oval, irregularly sinuously toothed at margin; lateral lobes 1 or several pairs, ovate. Petiole is wide, 2–6 cm long, amplexicaul at base. Lower cauline leaves, pinnately cleft, 6–10 cm long, expanded and amplexicaul at base, hispid and ciliated on both surfaces; upper cauline leaves oblong obovate, oblong or oblong lanceolate, 2.5–8 (15) cm long and 0.5–4 (5) cm wide, cordate and amplexicaul at base, pendulously auriculate on both sides, entire or undulately serrated at margin. Racemes are corymbose at anthesis, elongate afterwards. Flowers are bright yellow, 7–10 mm in diameter. Sepals are oblong, 3–5 mm long, erect, rounded at apex transparent at margin, slightly pubescent. Petals are obovate, 7–9 mm long, subretuse at apex, clawed at base. Silique is linear, 3–8 cm long, 2–4 mm wide; petals midribbed, reticulate; calyx erect, 9–24 mm long. Pedicel 5–15 mm long. Seeds are purplish brown, globular, 1.5 mm in diameter. Flowering is from March to April and fruiting in May.

Distributed in: Pohnpei, Yap, Chuuk, Kosrae

Joseph's Coat

Latin Name: ***Amaranthus tricolor***

English Name: Joseph's coat

Annual herbs, 80–150 cm tall. Stems are stout, green or red, often branched, hairy or glabrous when young. Leaf blades are ovate, rhombic-ovate or lanceolate, 4–10 cm long and 2–7 cm wide, green, red, purple, yellow, or partly green in mixture of other colors, obtuse or concave at apex, convex, cuneate at base, entire or undulate at margin, glabrous. Petiole is 2–6 cm long, green or red. Flowers are clustered, axillary up to lower leaves, or terminally clustered at the same time, forming drooping spikes. Flower clusters are globose, 5–15 mm in diameter, with a mixture of male and female flowers. Bracts and bracteoles are ovate-lanceolate, 2.5–3 mm long, transparent, 1 long aristate at apex, abaxially with 1 green or red raised midrib. Perianth segments are oblong, 3–4 mm long, green or yellowish-green, with a long awn tip at apex, abaxially with a green or purple raised midrib, Stamens are longer or shorter than perianth segments. Utricles are ovate-oblong, 2–2.5 mm long, circumscissile, enclosed in persistent perianth segments. Seeds are subrounded or obovate, 1 mm in diameter, black or black-brown, obtuse at margin. Flowering is from May to August and fruiting from July to September.

Distributed in: Pohnpei, Yap

Leaf Mustard

Latin Name: ***Brassica juncea*** **L.**

English Name: Leaf mustard

Annual herb, 30–150 cm tall, often glabrous, sometimes with spiny hairs on young stems and leaves, pruinose, spicy. Stems are erect, branched. Basal leaves are broad ovate to obovate, 15–35 cm long, obtuse at apex, cuneate at base, lyrate pinnatisect, with 2–3 pairs of lobes or not lobed, scalloped or toothed at margin; petiole 3–9 cm long, small lobed. Lower cauline leaves are small, scalloped or toothed at margin, sometimes obtusely toothed at margin, not ampexicaul. Upper cauline leaves are narrow lanceolate, 2.5–5 cm long, 4–9 mm wide, inconspicuously sparsely toothed or entire at margin. Racemes are terminal, prolonged after anthesis. Flowers are yellow, 7–10 mm in diameter. Pedicels are 4–9 mm long. Sepals are erect, pale yellow, oblong elliptic, 4–5 mm long. Petals are obovate, 8–10 mm long. Schizocarp is linear, 3–5.5 cm long, 2–3.5 mm wide, with 1 prominent midrib on petal; beak 6–12 mm long; pedicel 5–15 mm long. Seeds are purple-brown, 1 mm in diameter. Flowering is from March to May and fruiting from May to June.

Distributed in: Pohnpei, Yap, Chuuk, Kosrae

Common bean

Latin Name: ***Phaseolus vulgaris*** **L.**

English Name: Lima bean, Sierra bean

Annual herbs, twining or suberect. Stems are pubescent or glabrescent when old. Pinnate compound leaves have 3 leaflets. Stipules are lanceolate, 4 mm long, basal. Leaflets are broadly ovate or ovate-rhomboid, with lateral ones oblique, 4–16 cm long, 2.5–11 cm wide; apex long, acuminate, apiculate; base rounded or broadly cuneate, entire, pubescent. Racemes are shorter than leaves, several flowered at top of rachis. Pedicel is 5–8 mm long. Bracteoles are ovate, with several raised veins, persistent, approximately equal to or slightly longer than calyx. Calyx is cup-shaped, 3–4 mm long; upper 2 lobes combined into an emarginate lobe. Corolla is white, yellow, violet or red. Petal standard is nearly square, 9–12 mm wide; wing obovate; keel about 1 cm long, spirally twisted at apex. Ovary is pubescent; style flattened. Pods are banded, slightly curved, 10–15 cm long, 1–1.5 cm wide, slightly swollen, usually glabrous, apically beaked. Seeds are 4–6, oval or reniform, 0.9–2 cm long, 0.3–1.2 cm wide, white, brown, blue or variegated. Hilum is usually white. Flowering is in spring and summer.

Distributed in: Pohnpei, Yap

Cowpea

Latin Name: ***Vigna unguiculata*** **(Linn.) Walp**

English Name: Cherry bean

Annual herbaceous legume, twining or suberect, sometimes apically twining. Stems are almost subglabrous. Pinnate compound leaves have 3 leaflets. Stipules are lanceolate, 1 cm long, with a narrow spur on the point of attachment, striated. Leaflets are glabrous, ovate-rhomboid, 5–15 cm long, 4–6 cm wide, acute at apex, entire or nearly entire at margin, sometimes pale purple. Racemes are axillary, long pedunculate, 2–6 flowered. Flowers are clustered at top of rachis, often with dense fleshy glands between pedicels. Calyx is pale green, campanulate, 6–10 mm long; teeth lanceolate. Corolla is yellow-white, tinged with purple, about 2 cm long. Petals are petiolate; standard oblate, 2 cm wide, retuse at apex, slightly auriculate at base; wing slightly triangular; keel slightly curved. Ovary is linear, hairy. Pods are pendulous, erect or oblique, linear, 7.5–70 (90) cm long, 6–10 mm wide, slightly fleshy, inflated or solid, with many seeds. Seeds are oval or cylindrical or slightly reniform, 6–12 mm long, yellow-white, dark red or with other colors. Flowering is from May to August.

Distributed in: Pohnpei, Yap, Chuuk, Kosrae

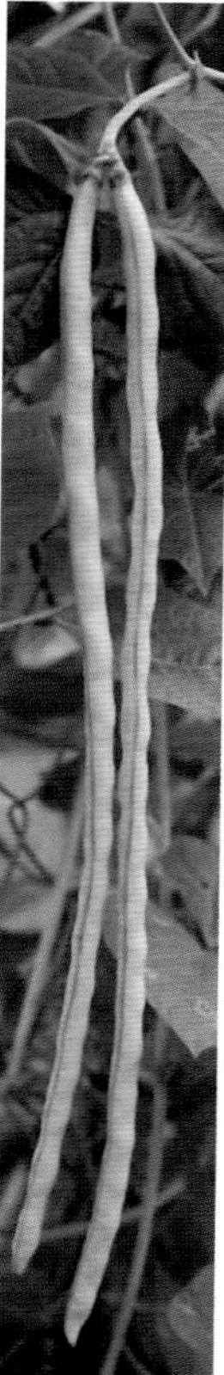

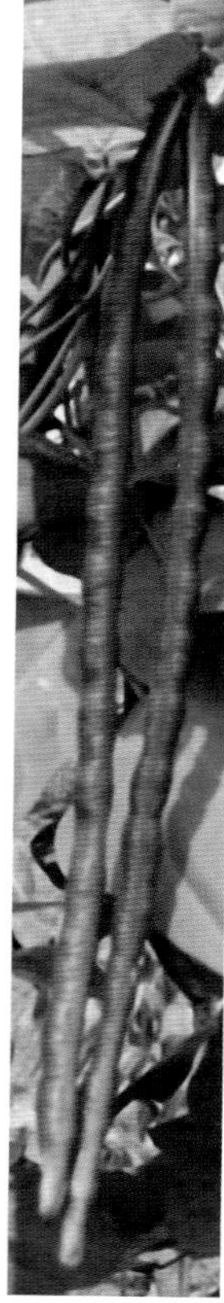

Sweet Potato Leaves

Latin Name: ***Ipomoea batatas*** **(L.) Lam.**

English Name: Sweet potato leaves

Annual tuberous herb. Stems are prostrate or ascending, occasionally twining. Leaves are usually broadly ovate, 4–13 cm long, 3–13 cm wide, entire or palmately 3–5 lobed. Petioles vary in length, 2.5–20 cm long. Cymes are axillary, 1–3(7) flowered, umbelliform. Bracts are small, lanceolate, 2–4 mm long, aristate or cuspidate at apex, caducous. Pedicel is 2–10 mm long. Sepals are oblong, acute or cuspidate at apex; outer sepals 7–10 mm long; inner sepals 8–11 mm long. Corolla is pink, white, pale purple or purple, campanulate or funnelform, 3–4 cm long. Stamens and style are included. Filaments are pubescent at base. Ovary is 2-4-loculed, pubescent, sometimes glabrous. Capsule is ovate or oblate. Seeds are 1–4.

Distributed in: Pohnpei, Yap, Chuuk, Kosrae

Water Spinach

Latin Name: ***Ipomoea aquatica*** **Forsk.**

English Name: Water spinach

Annual herbs, sprawling, terrestrial or aquatic. Stems are cylindrical, nodular, hollow at internode, rooted at node, glabrous. Leaf blades are variable, ovate, oblong ovate, oblong lanceolate or lanceolate, 3.5–17 cm long and 0.9–8.5 cm wide, subglabrous or occasionally sparsely pubescent on both surfaces; apex acute or acuminate, mucronate; base cordate, hastate or sagittate, occasionally truncated; margin entire, undulatey, or sometimes a few coarsely dentate at base. Petiole is 3–14 cm long, glabrous. Cymes are axillary. Peduncles are 1.5–9 cm long, pubescent at base, glabrous above base, 1-3-(5)- flowered. Bracts are squamiform (scale-like), 1.5–2 mm long. Pedicels are 1.5–5 cm long, glabrous. Sepals are subequal, ovate, 7–8 mm long, obtuse and mucronate at apex, abaxially glabrous. Corolla is white, reddish or purple, funnelform, 3.5–5 cm long. Stamens are unequal. Filaments are pubescent at base. Ovary is conical, glabrous. Capsules are ovoid to globose, 1 cm in diameter, glabrous. Seeds are densely pubescent, sometimes glabrous.

Distributed in: Pohnpei, Yap, Chuuk, Kosrae

Coriander

Latin Name: ***Coriandrum sativum*** **L.**

English Name: Coriander

Annual or biennial herb with a strong aroma, 20–100 cm tall. Roots are spindle-shaped, slender, with many slender lateral roots. Stems are terete, erect, multi-branched, striated, usually smooth. Basal leaves are petiolate. Petioles are 2–8 cm long. Leaf blades are 1 or 2–pinnatisect; pinnae broadly ovate or fan-shaped, 1–2 cm long and 1–1.5 cm wide, obtusely serrated, notched or parted at margin; upper cauline leaves are 3- or multi-pinnate; ultimate pinnae narrowly linear, 5–10 mm long, 0.5–1 mm wide, obtuse and entire at apex. Umbel is terminal or opposite to leaf; peduncle 2–8 cm long; rays 3–7, 1–2.5 cm long. Bracteoles are 2–5, linear, entire. Umbellules are 3–9 fertile flowered, white or tinged with purple. Calyx teeth are usually unequal; small ones ovate triangular; large ones oblong ovate. Petals are obovate, 1–1.2 mm long, about 1 mm wide, with concave ligule at apex; radial petals 2–3.5 mm long and 1–2 mm wide, usually entire, 3–5 veined. Filaments are 1–2 mm long. Anthers ovate, about 0.7 mm long. Styles are erect when young, flexed outward when mature. Fruit is spherical; dorsal main angles and adjacent secondary angles are conspicuous. Endosperm is ventrally concave. Oil tubes are not obvious, or only one beneath the secondary angle. Flowering and fruiting stages are from April to November.

Distributed in: Pohnpei, Yap

Ginger

Latin Name: ***Zingiber officinale*** **Rosc.**

English Name: Ginger

Plants are 0.5–1 m tall. Rhizome is thick, branched, aromatic, pungent. Leaf blades are lanceolate or linear lanceolate, 15–30 cm long, 2–2.5 cm wide, glabrous, sessile; ligule membranous, 2–4 mm long. Peduncles are 25 cm long. Inflorescence is a cone-shaped spike, 4–5 cm long. Bracts are ovate, about 2.5 cm long, pale green; sometimes yellowish at margin, mucronate at apex. Calyx tube is 1 cm long. Corolla is yellowish green; tubes 2–2.5 cm long; lobes lanceolate, less than 2 cm long. Central lobe of the labellum is oblong obovate, shorter than the corolla lobe, with purple stripes and yellowish spots; lateral lobes ovate, about 6 mm long. Stamens are dark purple. Anthers are about 9 mm long. Connetive appendage is subulate, about 7 mm long. Flowering is in Autumn.

Distributed in: Pohnpei, Yap, Chuuk, Kosrae

Spring Onion

Latin Name: ***Allium fistulosum*** **L.**

English Name: Spring onion, Welsh onion, long green onion

Bulbs are solitary, cylindrical, rarely turgic ovate-cylindrical at base, 1–2 cm wide, sometimes up to 4.5 cm wide. Tunic is white, rarely reddish brown, membranous to thinly leathery, not easy to break. Leaves are cylindrical, hollow, tapering to apex, subequal to scape, more than 0.5 cm wide. Scape is cylindrical, hollow, 30–50 (100) cm tall, swollen below the middle, narrowing to apex, sheathed beneath about 1/3 of scape length. Spathe is membranous, 2 valved. Umbel is globose, many flowered, sparsely arranged. Pedicels are slender, equal to perianth segments, or 2–3 times longer than perianth segments, ebracteolate at base. Perianths are white; segments 6–8.5 mm long, subovate, acuminate at apex, with a reflexed point, ebracteolate at base; inner ones are longer than the outer. Filaments are 1.5–2 times longer than perianth segments, subulate, connate at base, adnate to perianth segments. Ovary is obovate, with inconspicuous nectaries at base of ventral suture. Style is slender, protruding beyond perianth. Flowering and fruiting stages are from April to July.

Distributed in: Pohnpei, Yap, Chuuk, Kosrae

Garlic

Latin Name: ***Allium sativum*** **L.**

English Name: Garlic

Bulbs are globose to oblate, usually consisting of several fleshy, compactly arranged bulbels, covered with several white to purple membranous scales. Leaves are broad linear to linear lanceolate, flattened, long acuminate at apex, shorter than scape, 2.5 cm wide. Scape is solid, terete, up to 60 cm long, sheathed below the middle. Spathe is caducous, with a beak 7–20 cm long. Umbel consists of many bulbels and a few flowers. Pedicles are slender. Bracteols are large, ovate, membranous, acute at apex. Flowers are usually pale red. Perianth segments are lanceolate to ovate lanceolate, 3–4 mm long; inner segments shorter than outer ones. Filaments are shorter than perianth segments, connate at base, adnate to perianth segments; inner ones enlarged at base, 1 toothed on each side at base, filiform at tooth apex, with the tooth longer than the perianth segment; outer ones subulate. Ovary is globose. Styles are not projected beyond perianth. Flowering is in July.

Distributed in: Pohnpei, Yap, Chuuk, Kosrae

Onion

Latin Name: ***Allium cepa*** **L.**

English Name: Onion

Bulbs are thick, subglobular to oblate. Tunic is purple red, brown red, pale brown red, yellow to pale yellow, papery to thinly leathery. Inner scales are thick, fleshy, not easy to break. Leaves are cylindrical, fistulous, the thickest below the middle part, narrowed upward, shorter than scapes,more than 0.5 cm wide. Scapes are stout, up to 1 meter tall, fistulous, cylindrical, enlarged below middle, narrowed upward, covered with leaf sheaths at lower part. Spathe is 2–3 valved. Umbel is globose, densely many flowered. Pedicels are about 2.5 cm long. Pollen is white. Perianth segments are oblong ovate, 4–5 mm long, 2 mm wide, with green midrib. Filaments are equal, slightly longer than perianth segments, connate at about 1/5 of base, adnate to perianth segments at 1/2 of the connate part; inner ones extremely enlarged at base, 1 toothed on each side of the base; outer ones subulate. Ovary is globose, with concave nectary pit covered with hoodlike projections at base of ventral suture. Style is 4 mm long. Flowering and fruiting stages are from May to July.

Distributed in: Pohnpei, Yap, Chuuk, Kosrae